Information Circular 9505

Age Awareness Training for Miners

By William L. Porter, Launa G. Mallett, Ph.D., Diana J. Schwerha, Ph.D.,
Sean Gallagher, Ph.D., Janet Torma-Krajewski, Ph.D., and Lisa J. Steiner

DEPARTMENT OF HEALTH AND HUMAN SERVICES
Centers for Disease Control and Prevention
National Institute for Occupational Safety and Health
Pittsburgh Research Laboratory
Pittsburgh, PA

June 2008

This document is in the public domain and may be freely copied or reprinted.

Disclaimer

Mention of any company or product does not constitute endorsement by the National Institute for Occupational Safety and Health (NIOSH). In addition, citations to Web sites external to NIOSH do not constitute NIOSH endorsement of the sponsoring organizations or their programs or products. Furthermore, NIOSH is not responsible for the content of these Web sites.

In considering the recommendations contained in this report, employers need to be aware of federal laws, such as the Age Discrimination in Employment Act of 1967 (ADEA), as well as applicable state and local laws, that may impact their implementation.

Ordering Information

To receive documents or other information about occupational safety and health topics, contact NIOSH at

>Telephone: **1–800–CDC–INFO** (1–800–232–4636)
>TTY: 1–888–232–6348
>e-mail: cdcinfo@cdc.gov
>
>or visit the NIOSH Web site at **www.cdc.gov/niosh**.

For a monthly update on news at NIOSH, subscribe to NIOSH *eNews* by visiting **www.cdc.gov/niosh/eNews**.

DHHS (NIOSH) Publication No. 2008–133

June 2008

SAFER • HEALTHIER • PEOPLE™

ACKNOWLEDGMENTS

The authors thank the following NIOSH personnel: Scott Verlihay (Summer Intern), E. William Rossi (Industrial Engineering Technician (retired)), and Leigh McClure (SCEP Student) for developing documents, illustrations, and demonstrations; Robin Burgess-Limerick, Ph.D. (NAS Fellow and Associate Professor, University of Queensland, Australia), for content review; and Richard L. Unger (Safety Engineer) for assistance in creating graphics for the Web-based interface. The authors also thank Unimin Corp. and U.S. Silica Co. for their assistance in field testing the training package, with special thanks to Bob Newman, Phil Boyd, and the employees at Unimin's Gleason, TN, plant.

TABLE OF CONTENTS

Overview
    Introduction
    Training objectives
    Training content
    Uses of training modules
    Implementing AAT at a mine facility: an example
    Needed materials
    File formats

Instructor's Guide
    Who is involved with the training?
    How do I implement the training?
    What does the training include?

Age Awareness Training – Leader Level

Age Awareness Training – General Level
    Module 1: Introduction to Age Awareness
    Module 2: Vision
    Module 3: Hearing
    Module 4: Attention and Memory
    Module 5: Musculoskeletal System
    Module 6: Lower Back
    Module 7: Work Capacity and Endurance
    Module 8: Slips, Trips, and Falls
    Module 9: Identifying High-Risk Tasks

Age Awareness Training – Records / Evaluations

References

Glossary of Terms

Overview

Age Awareness Training

Introduction

From the first day of new miner training until the day they retire, mine workers will experience changes due to the normal aging process. It is an unfortunate fact of life that many age-related changes result in diminished physical, sensory, or cognitive capabilities. Of course, workers also gain a tremendous wealth of experience, knowledge, and insight as they age, making them a vitally important resource for their company. Effective leveraging of this precious resource requires both an appreciation of the changes that occur with age and an understanding of methods that can be used to reduce the injury risk that may result. The purpose of this training is to provide the information necessary to accomplish these objectives.

Aging workers may not necessarily have a higher injury risk overall; however, the effects of a musculoskeletal injury (MSI) on older workers may be more extreme. MSHA data show that not only does the percentage of MSIs increase when workers are over age 30, so does the number of days lost per injury, as shown in the graph below [MSHA 2005].

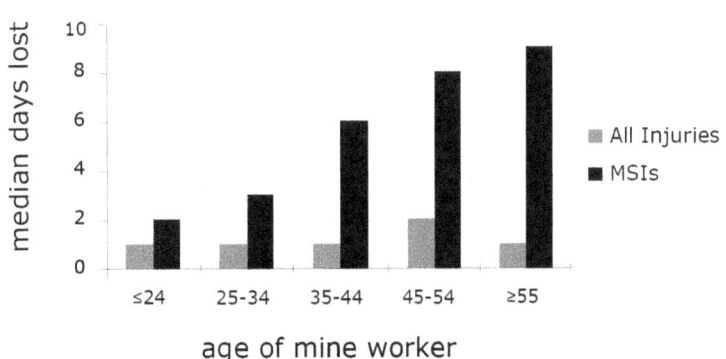

Protecting the safety and health of aging workers requires matching the demands of the job to worker capabilities. This means reducing or eliminating risk factors for injuries, such as heavy lifting, awkward postures, static postures, repetitive movements, and/or vibration exposures. In other cases, it may mean improving visibility or streamlining processes. Designing jobs to accommodate the changing capabilities of older workers will not only reduce injury risk for older workers, it will have the added benefit of protecting younger employees as well.

Training objectives

After completing this training, miners will:
- Better understand age-related changes that everyone experiences
- Identify work tasks and situations that put older workers at risk
- Be familiar with ways to modify jobs to accommodate older workers
- Know about lifestyle choices that can keep them healthier both on and off the job

Training content

- Presentation for program leaders (trainers, supervisors, labor representatives)
- Modules for use with all employees
    - Safety talk guides
    - Activity guides
    - Newsletters
- Evaluation/training record forms

Uses of training modules

This training package was designed to be flexible and work with your current training system. The modules can be given in a biweekly (weekly, monthly) series of tailgate safety talks. The material can also be used in other ways that may work better with a company's safety and health training program. Several modules can be given at one time as part of annual refresher training or quarterly safety meetings. You may also use individual modules to supplement current training programs.

There is no formal order for the training modules, except that the Introduction and the Risky Task modules should be given first and last, respectively. Use the modules however they best fit the needs of your company. For example, if you have recently had hearing conservation training, you may want to move the hearing module to the end of the training so it can serve as refresher training.

The Leader Level PowerPoint presentation in its current format is written for program leaders. This presentation can be modified with your company's demographic and injury data. This tailored information can be presented to management as justification for investing in the training package.

Implementing AAT at a mine facility: an example

The Age Awareness Training (AAT) program was tested at Unimin Corp.'s clay facility in Gleason, TN, during the summer and fall of 2006. The age of employees at this facility ranged from 22 to 68, with a median age of 45.5 years. This diversity of the workforce created a perfect field test location for the AAT. The first "kickoff" module (Introduction) was given at monthly safety meetings to all shifts over a period of 1 week to a total of about 45 employees. This was followed with biweekly safety talks given by the supervisors on each of the modules.

The Gleason Mine Safety and Health Supervisor commented: "Overall the vision module (the first to be given in the biweekly format) was received rather well, and has provoked some good discussion and a few projects." For example, one intervention resulting from the vision module was applying antiglare film on some mobile equipment windows. The effect of glare on an aging population was defined in the vision module. Given that older workers may have problems adapting to situations with glare, adding antiglare film reduced the risk of injury to the workers by allowing them to be able to distinguish possible hazards. Gleason employees made many other workplace modifications as a result of the training. These included:

- Vision
    - Added lighting in low-light and transition areas identified by workers. (Most supervisors work on day shift and never see low-light areas. The supervisor prior to training did not see this as a problem, but workers identified it right away.)
    - Added tinted film on equipment in mine pit to reduce glare
- Hearing
    - Cleaned door seals on equipment
    - Repaired equipment that produced high noise levels
- Slips, trips, and falls
    - Allowed for drainage of walkways and standing water around site
- Musculoskeletal
    - Controlled spillage at problem areas around conveyor belts (eliminated the need for workers to clean up spillage)
- Back
    - Gas cylinders delivered directly to storage area. (Before training, the gas company would drop off cylinders near the office, and workers would "manhandle" them to a storage area. Now the gas company delivers the cylinders to the storage area since they have the equipment to properly handle them.)
- Attention/Memory
    - Made sure everyone has the resources and places to keep checklists (reinforced by management to support making checklists)

Initially, Unimin was concerned that the training package may be perceived as age-discriminatory. After discussion with NIOSH researchers regarding the specific aims of the training package, the company decided to proceed with the training. Upon completion, Unimin's General Manager of Safety and Health commented that the company and the employees were pleased with how the training was handled. Unimin did not feel that the training stigmatizes older workers, stating that "the lessons are suitable and appropriate for miners/workers of all ages." The Safety and Health Supervisor at Gleason, upon completing the training, did inform NIOSH researchers that the training was a "harder sell for the younger guys," although he stated that some of the younger workers "really latched on to it." Workers aged 40 and older were very interested in the training topics.

The Safety and Health Supervisor did say that both the back and the hearing modules were somewhat redundant to information that Unimin had already presented in other training packages for hearing conservation and proper lifting techniques. Although this material may be somewhat redundant, it presents the topics in a different way by examining the aging aspect of the issues, and it still produced workplace improvements. As noted above, Gleason made improvements based on information received from all of the modules. The company reported that some of the changes had been considered before, but the training highlighted the reasons for and importance of such changes and made everyone understand that making them should be a priority. Unimin Corp., mine management, and, most importantly, the workers were pleased with the results of the training, and Unimin plans to expand the training to other facilities in the future.

Needed materials

To complete all of the training modules, the following materials are required. The materials needed for each section of training are detailed in each module.

General Materials:
- Projector or other means of displaying a PowerPoint presentation (needed to present the Leader Level Training)
- Color printer or color computer monitor needed to display activity graphics
- Photocopier (to make copies of safety talks and newsletters)

Material Required for Module Activities:
- Microsoft Windows-capable computer with speakers
- A light weight (3–5 pounds)
- Stopwatch or watch with time in seconds
- Paper clip or a metal coat hanger
- NIOSH Hearing Loss Simulator software (downloadable from http://www.cdc.gov/niosh/mining/products/product47.htm)
- Nine 3/16- by 5-inch wooden dowels
- Nine 1/4- by 5-inch wooden dowels
- One 1-inch round polyvinyl chloride (PVC) pipe
- One 1/2-inch round shrink tubing
- Optional: a calculator and a heart rate monitor

File formats

This Information Circular (IC) is available in hard-copy format, electronically on CD, and is downloadable from the NIOSH Mining Web site (http://www.cdc.gov/niosh/mining/).

Electronic Versions

Each portion of the training is provided in Microsoft Word or PowerPoint format and as a PDF. The Microsoft Word files may be modified to include specific examples about your site that may be more meaningful to your employees. The Microsoft PowerPoint file may be tailored to the specific need of your company (if you do not have a copy of Microsoft PowerPoint, a free viewer can be downloaded from www.microsoft.com/downloads/). The PDF versions of all documents are included if you do not have access to Microsoft's software. A Web-based document management system is used in managing these documents. This system should automatically begin when the disk is inserted into the computer. If this does not occur, start by opening the file called "index.html". The document management system is compatible with Microsoft Internet Explorer 6 or greater, Mozilla Firefox 1.5 or greater, and Netscape 8 or greater. The following table of commands is used for the Web-based document system.

Age Awareness Training document management system navigation

COMMAND BUTTON	ACTION
⊙ (forward arrow)	Move forward in training module
⊙ (back arrow)	Go back to previous page in module
●	Go to specific section
T	Shortcut to General Level Topics menu
PDF	Open documents for section in PDF format
Training Objectives	Displays key training points for the module
Contact Us	Displays contact information for program support
Preview	Shows version of the Leader Level PowerPoint presentation viewable with PowerPoint software
EXIT	Exit the training module

Computer Tips

Information Bar Warning Viewing Contents in Internet Explorer 7 (IE7)

To help protect your security, some computers have restricted this Web page from running. You can choose to permit the content to load by clicking "OK" on the Information Bar displayed by Internet Explorer.

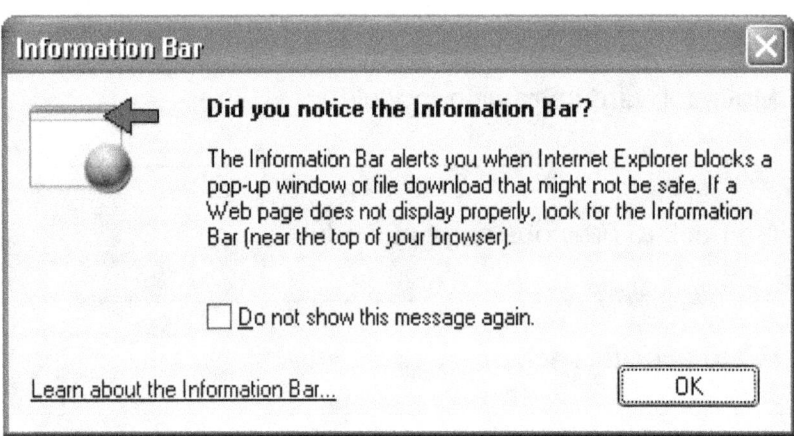

Possible Error

In IE7, if you see the following error message "Page pxx does not exist and could not be loaded," go to "Tools, Internet Options, Advanced Tab." Scroll down to the Security section and clear the "enable native XMLHTTP support" option.

Disabling the Image Toolbar

You may have noticed that whenever your mouse pointer moves over an image, an image toolbar automatically pops up, giving you the option to save, print, or e-mail a picture.

Disable the Image Toolbar by Using the Image Toolbar:

1. Place your mouse cursor over an image. When the image toolbar appears, right-click it, and then click "Disable Image Toolbar."
2. Click "Always" to disable the image toolbar for all future browsing sessions, or click "This Session" to disable the image toolbar for the current session only.

Age Awareness Training
Instructor's Guide

Whether your workers average 25 or 55 years old, they are aging, and this impacts safety and health at your operation. This training was designed to teach which normal changes happen as people age, ways that people can protect themselves from age-related safety and health risks, how age-related changes can impact ability to work, and workplace adaptations to fit different employee ages.

If you have an older workforce, you may have already seen the impact that years of working have had on some employees. Perhaps you are seeing a rise in sprains and strains. If you have a younger workforce, you may be seeing employees with limited experience on the job. Now is the time to teach them to protect themselves as they work so they will remain safe and healthy for many years to come.

There are two levels of training provided. The first is for your leaders and is called Leader Level Training. Managers, safety committee representatives, front-line supervisors, and whoever conducts safety talks should be given the Leader Level Training. This training helps the leaders prepare for the employee training, which is called the General Level Training. All employees should be involved in the General Level Training. It can be given as part of your routine safety talk program. Don't forget to include sessions for your lab and/or office staff. They are aging, too.

Get started by reading this Instructor's Guide. It will give you an overview of the training and details about implementation. Then read through the various segments of the Leader Level and General Level Training. There is opportunity for customization so you can make the training fit your situation.

Who is involved with the training?

There are three main roles for people participating in the training, which are outlined below. Select individuals or groups for each role.

The Champion

- Develops an implementation plan for the training
- Manages the training implementation
- Gives Leader Level Training to identified site leaders
- May conduct all the training or may train leaders to give some portions
- Is responsible for reviewing recommendations from employees that are developed as a result of the training and ensuring that these are considered for implementation
- Evaluates the training

Leaders

- Are advocates for the importance of this training
- Attend Leader Level Training
- Present General Level Training segments as designated in the implementation plan (any leader may give one or more modules to one or more groups of employees)
- Work with the champion to manage suggestions that are made as a result of the training
- Work with supervisors to review suggestions given during safety talks and implement them when possible and appropriate
- Evaluate how the participants received and learned from the material in each module
- Provide feedback to employees regarding implemented changes

Trainees

- Participate in General Level Training
- Review their work tasks and environments to look for age-related risks
- Make recommendations for improvements based on what they have learned
- Protect themselves with healthy lifestyle choices learned during the training

All three roles are important for successful training. The champion must believe in the training and approach it with enthusiasm. The leaders must be ready to "spread the word" about the importance of the training and its potential impact. Everyone must be open to looking at work tasks, procedures, and designs with an eye to improving them for older workers and for younger workers who want to age safely.

How do I implement the training?

There are four phases to implementing this training package. This section provides an overview of these phases and some detail about the material covered.

Phase 1: Planning for Training (about 2 weeks)
Phase 2: Training Leaders (about 2 weeks)
Phase 3: Training Workers (about 2–9 months)
Phase 4: Postwork (ongoing)

Phase 1: Planning for Training

The first step is to establish who will be the champion and the leaders. The champion should work with leaders and discuss the general principles of the training and the time commitment needed to properly complete the training. The champion, using feedback from the leaders, should develop an implementation plan by setting a schedule for the training, determining the order of the modules that will be presented, and deciding how the newsletters will be distributed (i.e., distributed at a safety talk, sent home to employees as a mailing, or in a company newsletter). A strategy for evaluating the training should also be planned.

Phase 2: Training Leaders

The next step in implementation is to give the Leader Level Training to all designated leaders. The champion will conduct this classroom-based session with the materials provided. A PowerPoint presentation with accompanying notes can be used as is or modified for a given situation.

After the champion presents the Leader Level Training, the leaders responsible for giving General Level Training session should review the material associated with their modules. They should start by reading the newsletter for their module(s). The accompanying safety talk guides are written in such a way that the leader may choose to read the guide(s) word for word or develop his or her own script based on the material provided. Leaders should also prepare to conduct the activities associated with the module. When preparing for a safety talk, the leader should be ready to discuss the questions that are listed on the second page of the safety talk, write down trainee responses, and report the responses back to the champion. The leader can also review the additional resources included within each module. Leaders will also need to decide if they want to use the activities associated with the module. If they choose to use them, they will need to prepare any necessary resources.

Phase 3: Training Workers

The General Level Training phase may occur in several ways depending on the implementation plan developed in Phase 1. Each module of the General Level Training includes a newsletter on the specified topic, a safety talk guide, related activities, and additional resources. The newsletters can be distributed as part of a company newsletter, in paycheck envelopes, or during safety meetings. The safety talks can be presented in many ways, e.g., to small groups (10–15) of workers during a biweekly meeting or monthly during a company safety meeting that is attended by all workers. As an alternative to using all of the modules, selected modules may be given as part of another training plan, such as annual refresher training. The number and timing of modules presented should be planned to match individual company needs.

The Introduction module should be the first module given. The champion may conduct this training if he/she wants it to serve as the "kickoff." A kickoff meeting is a good opportunity to generate employee enthusiasm by showing management support for the training. Alternatively, each individual leader may give the introductory module to small groups.

After the Introduction module is given, the topic-specific modules should be presented. The safety talks are designed to take less than 15 minutes and to be toolbox-style talks. During the talks, trainees will be asked to make initial assessments of risks in their work areas related to the topic of the talk. They should be encouraged to later identify risks with their supervisors, the leader giving the talk, or the Age Awareness Training champion. The topic modules are the core of the training package.

Once the other training modules have been completed, the module on Identifying High-Risk Tasks should be given. This module provides scenarios describing work tasks in various environments. The leaders will read the scenario to the trainees and ask them to identify potential risk factors that may be affected by age-related changes and suggest changes that could reduce the risk of injury. This step is an important tool in determining how successful the safety talks have been at getting the trainees to think about age-related issues.

Phase 4: Postwork

Some type of evaluation should occur for the training package. The type of evaluation used should be included in the implementation plan developed by the champion prior to starting the training. Each module can be informally evaluated by the leaders themselves and/or the champion may evaluate the training using the evaluation forms included with the package. This step is important in establishing the effectiveness of the training. If evaluations occur, then the champion will have a way to track if the training was a success or if the material was not well received. Either way, this information is important. If the training was not well received, then the champion can figure out ways to make the training more interesting to the trainees. If the training was a success, then the champion will have a way of documenting the success and may want to provide that information to management to validate the investment in this training and future training packages.

The material covered in this training should be revisited 6–12 months after the training is completed.

What does the training include?

This training was designed to be very flexible for the leader. Each portion of the training is included in Microsoft Word or PowerPoint format and as a PDF. The Microsoft Word files may be modified to include specific examples about your site that may be more meaningful to your employees. The Microsoft PowerPoint file may be tailored to the specific need of your company. (If you do not have a copy of Microsoft PowerPoint, a free viewer can be downloaded from www.microsoft.com/downloads/). The PDF versions of all documents are included if you do not have access to Microsoft's software. Along with these documents, a sticker design is included in the package. This design or one generated by your company is a useful tool to develop awareness about the training and remind workers about the training after it has ended.

Specific Contents for Phase 2:

Leader Level Training

This is a PowerPoint presentation that is designed as an overview of the training package for the leaders and management. *The presentation is not designed to be a stand-alone training tool since it only introduces the issues and does not discuss how they could affect the workers' health and safety.* It can be modified by including company-specific statistics, examples, photographs, or goals for injury reduction.

The champion may want to give this presentation formally to leaders. Alternatively, if the champion will be presenting all of the General Level Training him/herself, the presentation may be given to management/supervisors to get their feedback on the training package. This will help them prepare to review risks identified by employees during training and answer questions they may be asked. The PowerPoint presentation will be useful background material for the champion as he or she prepares to give General Level Training.

Specific Contents for Phase 3:

General Level Training

Each module listed below contains a safety talk guide, a newsletter, one or more activities, and additional resources. The safety talk guide is a two-page guide for the leader that contains major points about the topic and questions for the employees regarding how the topic relates to their workplace or health. The safety talks are designed to be given in 10- to 15-minute time periods and can be done in the field. The newsletter provides information about normal age-related changes, changes that can be made to the workplace, and activities the employee can do to help reduce the effects of aging. The newsletter can be distributed to the employees at the same time as the safety talk, sent home to the employees as an individual mailing, or included in a company newsletter. The leader may want to conduct the activity as a demonstration of the principles covered during the talk. Most activities can be done in the field without the aid of a computer. The modules also have additional resources and refer to relevant regulations. These are not intended to be given to all employees, but are resources for the leader if more information is needed.

<u>Module 1: Introduction to Age Awareness</u>

People go through normal physiological changes as they age. This section introduces key concepts and discusses why workers should be aware of normal age-related changes. Participants will learn about why normal age-related changes may affect their safety, health, and work performance.

Module 2: Vision

Most people will go through some normal age-related vision changes during their lifetime. This section provides information on how these changes occur and how to improve the workplace to accommodate age-related vision changes. Participants will learn about what normal age-related changes may occur to their vision and how they can modify the work environment to accommodate those changes.

Module 3: Hearing

Many miners experience hearing loss throughout their working life. This hearing loss can be age-related, noise-induced, or some combination of both. This section provides information as to why people experience hearing loss, how people can protect their hearing, how the workplace can be modified to accommodate those changes, and how people can protect themselves from hearing loss risks.

Module 4: Attention and Memory

People experience normal age-related changes to their attention and memory that may affect how they respond to people, signals, and events. This section provides information on how workers can modify their work environment so that these types of changes will not affect their safety and performance. Participants will learn which elements of memory and attention are affected by aging and ways to modify the work environment to match work demands with worker capabilities.

Module 5: Musculoskeletal System

Strain and sprain injuries are common experiences in mining and tend to occur more frequently with age. This section provides information as to how and why these injuries occur and gives recommendations to reduce sprain/strain injury risks. Participants will learn how the musculoskeletal system changes as they age and how they can modify their work environment to accommodate changes and maintain their musculoskeletal health.

Module 6: Lower Back

Low back pain is one of the leading causes of disability in the mining industry. This section describes how back pain develops, the role of age, and provides advice on how the risk of back pain can be reduced. Participants will learn how low back pain can develop as they age and how they can reduce back injury risk by modifying the work environment or improving individual behaviors.

Module 7: Work Capacity and Endurance

Aging causes a gradual reduction in the amount of physical effort that workers can sustain without fatigue. This section includes recommendations on how workers can reduce their workload to prevent physical overload, which may lead to serious injuries. Participants will learn how reduced work capacity can lead workers to make mistakes or become injured and ways they can modify the worksite to accommodate these normal age-related changes.

Module 8: Slips, Trips, and Falls

A significant number of injuries sustained by mine workers are attributed to slips, trips, and falls. This section presents information on why older workers tend to have a higher incidence and severity rates, and offers advice on how to reduce the risk of falls around the worksite. Participants will learn why slips, trips, and falls affect older workers to a greater degree and how they can modify their work environment to reduce their risk of injury.

Module 9: Identifying High-Risk Tasks

This section provides scenarios of work tasks to the miners. The scenarios include aspects of all of the normal age-related changes that are covered in the modules. Participants will read the scenarios and identify those areas that pose a risk of injury to an age-diverse workforce. Participants should be able to identify at least two risks for each scenario.

Specific Content for Phase 4:

Records/Evaluation Forms

Each module should be evaluated by the leaders presenting the General Level Training. The leaders can evaluate the effectiveness of the modules by reviewing the answers discussed during training to the questions on the second page of the safety talk. If the trainees were not able to generate discussion for these questions, the module could be revisited with additional examples. The participants should also be asked what they thought of each module.

The champion may also choose to evaluate the training by using the two included evaluation forms. This will also provide a way to keep a record of how and when the training was conducted. The Leader Level Training evaluation form is designed for use by the champion to keep a record of the leader training in Phase 2 of the training package. The General Level Training evaluation form is designed to get leader feedback on how each module was given and how it was received. The leader(s) and champions should also review the risks identified and the suggestions made by employees during the safety talks. These evaluations will help show whether employees have understood the concepts and can apply them to their workplaces. Risks identified must be reviewed to see if changes need to be made, and suggestions must be reviewed to show that management is committed to the training. Communications to provide feedback to employees after these reviews should be part of the implementation plan.

Another way to evaluate the effectiveness of the training package is by using the activity for the Introduction module. This survey is a way to determine the trainees' thoughts about aging and about how it affects them. This can be used as a pretest (given before training begins) and a posttest (given after training ends). The champion can then compare the results of the two surveys to determine if the training changed the way the trainees think about aging and its relationship to their work.

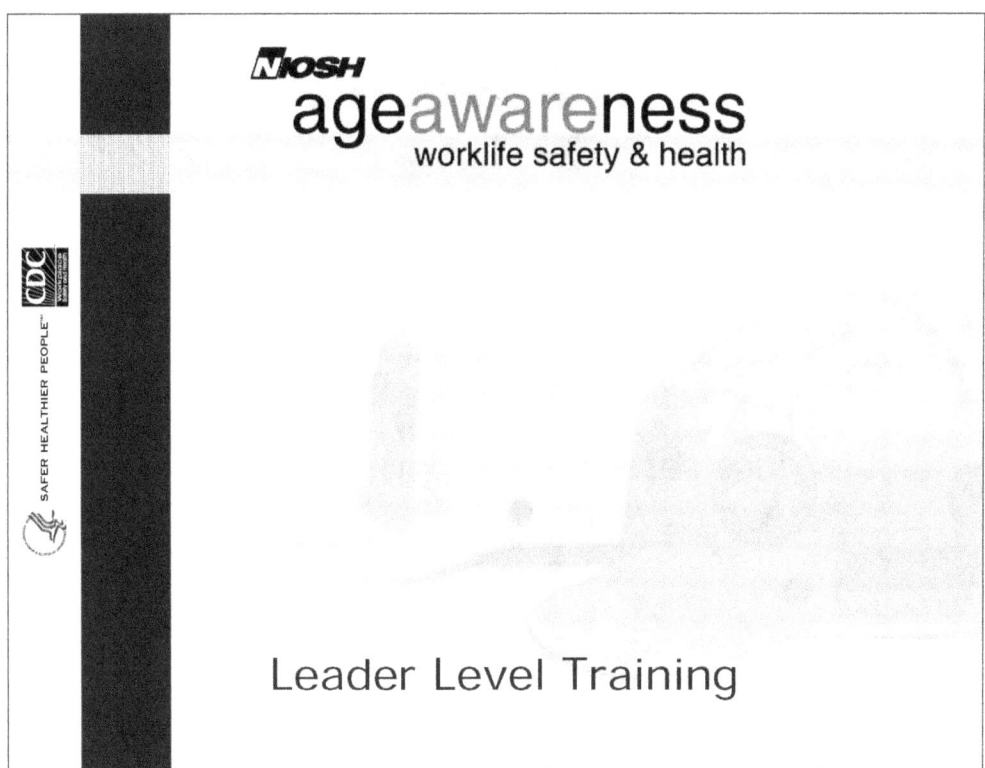

NOTE: This Leader Level presentation provides an overview of the Age Awareness Training (AAT) for organization leaders. Please review the Instructor's Guide for guidance on the use and implementation of this training presentation.

Key Points:

This presentation is an overview of the Age Awareness Training (AAT). It is set up in four parts:
- Myth or Truth?
- The Real Issues
- Addressing the Truths
- Training Topics

You may go through this presentation alone as you prepare to be the AAT champion and/or trainer at your site.

Depending on how delivery of the AAT is structured at your site, you may also want to present this material to other leaders and trainers. Before giving the presentation you may choose to:
- Shorten or extend the training in certain areas.
- Add some of your own statistics or goals for injury reduction.
- Modify it with your own pictures or videos.

Is It True?

There are "facts" we all know about aging.

But does research support what we "know"?

Key Points:

- To start the training, we will dispel some myths that are sometimes associated with aging.
- When we discuss aging, we are not only talking about workers over 55; we are discussing age-related changes that start as early as in your twenties.
- Whether you are 30 or 55 years old, you are aging and this impacts your safety and health.

> # Myth or Truth?
>
> "You can't teach an old dog new tricks."
>
> - Older workers can learn new tasks.
> - They may take longer to master something new, but they will become just as skilled as younger workers once they've "got it."

Key Points:

- Older workers do benefit from training.
- Older people can achieve the same level of competency as younger ones.
- However, younger ones learn new tasks more quickly.

Slide Animations:

- At first, the title "Myth or Truth?" and a quoted statement will be on the screen.
- Click for the green "no" symbol and supporting evidence to appear.

Activity Instructions:

- When you first come to the slide (only the quoted statement will appear), ask the audience members if they feel the statement is true or false and why.
- Advance the slide (so that the green "no" symbol appears), and discuss the statements made by the audience.

Myth or Truth?

"It's not cost-efficient to spend money training older workers."

- Older workers stay on their jobs after training as long as younger workers.
- Training for younger *and* older workers is a good investment.

Key Points:

- Even though it may seem a better investment to train younger employees, this isn't always so.
- Older workers tend to stick with the same job and company; therefore, they have lower turnover rates.
- Training may keep older workers more interested in their jobs.
- On the other hand, providing training for younger workers is a benefit that attracts the best employees and shows an investment by the company in the employees' safety and health.

Slide Animations:

- At first, the title "Myth or Truth?" and a quoted statement will be on the screen.
- Click for the green "no" symbol and supporting evidence to appear.

Activity Instructions:

- When you first come to the slide (only the quoted statement will appear), ask the audience members if they feel the statement is true or false and why.
- Advance the slide (such that the green "no" symbol appears), and discuss the statements made by the audience.

Myth or Truth?

"Older workers drive up workers' compensation costs."

- When injured, older workers recover more slowly than younger ones

But...

- Younger workers have higher injury rates

So...

- Neither causes higher overall costs.

Key Points:

- When injured, older workers generally need more days to recover than younger ones, but they have less frequent injuries.
- This means return-to-work programs can be especially useful for an aging workforce.
- Younger workers are generally injured more often, but they typically recover more quickly.
- Because these balance out, neither causes higher overall costs.

Slide Animations:

- At first, the title "Myth or Truth?" and a quoted statement will be on the screen.
- Click for the green "no" symbol and supporting evidence to appear.

Activity Instructions:

- When you first come to the slide (only the quoted statement will appear), ask the audience members if they feel the statement is true or false and why.
- Advance the slide (such that the green "no" symbol appears), and discuss the statements made by the audience.

Myth or Truth?

"As they age, workers become less productive."

- Older workers do the job "smarter."
- Younger workers have higher absentee rates.

Key Points:

- Research has found no data to support this.
- Older workers often increase their productivity by finding "smarter" ways to accomplish demanding tasks.
- Even though younger workers are generally healthier, they tend to miss more work, which reduces their productivity.

Slide Animations:

- At first, the title "Myth or Truth?" and a quoted statement will be on the screen.
- Click for the green "no" symbol and supporting evidence to appear.

Activity Instructions:

- When you first come to the slide (only the quoted statement will appear), ask the audience members if they feel the statement is true or false and why.
- Advance the slide (such that the green "no" symbol appears), and discuss the statements made by the audience.

Myth or Truth?

"It's too late for older workers to adopt healthy lifestyles."

- All workers benefit from healthy choices:
  - Tobacco cessation
  - Weight management
  - Formal wellness program participation
- Injury and illness prevention is the goal.

Key Points:

- As we age, our health is not only a result of our current health choices, but also the choices we've made over our entire lifetime. This is why it is important for people of all ages to adopt healthy choices.
- It is never late—or too soon—to start improving your health habits.
- It is much better to prevent injury and illness than to remedy problems after they occur.

Slide Animations:

- At first, the title "Myth or Truth?" and a quoted statement will be on the screen.
- Click for the green "no" symbol and supporting evidence to appear.

Activity Instructions:

- When you first come to the slide (only the quoted statement will appear), ask the audience members if they feel the statement is true or false and why.
- Advance the slide (such that the green "no" symbol appears), and discuss the statements made by the audience.

The Real Issues

Normal age-related changes affect all workers.

Key Points:

- Now that we have dispelled the myths surrounding aging, let's look at truths of normal age-related changes.

The Aging U.S. Workforce

- 40.7% of workforce is 45 or older.
- 52.3% of mining workforce is 45 or older.

Median Ages in 2006

- 40.8 – workforce
- 43.2 – metal
- 46.6 – coal
- 44.0 – nonmetal and quarry

Key Points:

- The U.S. workforce is aging.
- 40.7% of the U.S. workforce is 45 or older.
- This trend is even more prominent in mining, where 52.3% of workforce is 45 or older.

- NOTE: The *mean* of a set of numbers is found by dividing the sum of the data by the number of entries (it is also called the average). For injury statistics, we often use median as a way of describing the data. The *median* is the point at which half of the data is smaller and half of the data is larger. For data that may have several extreme points, such as injury data, medians have more meaning than averages.

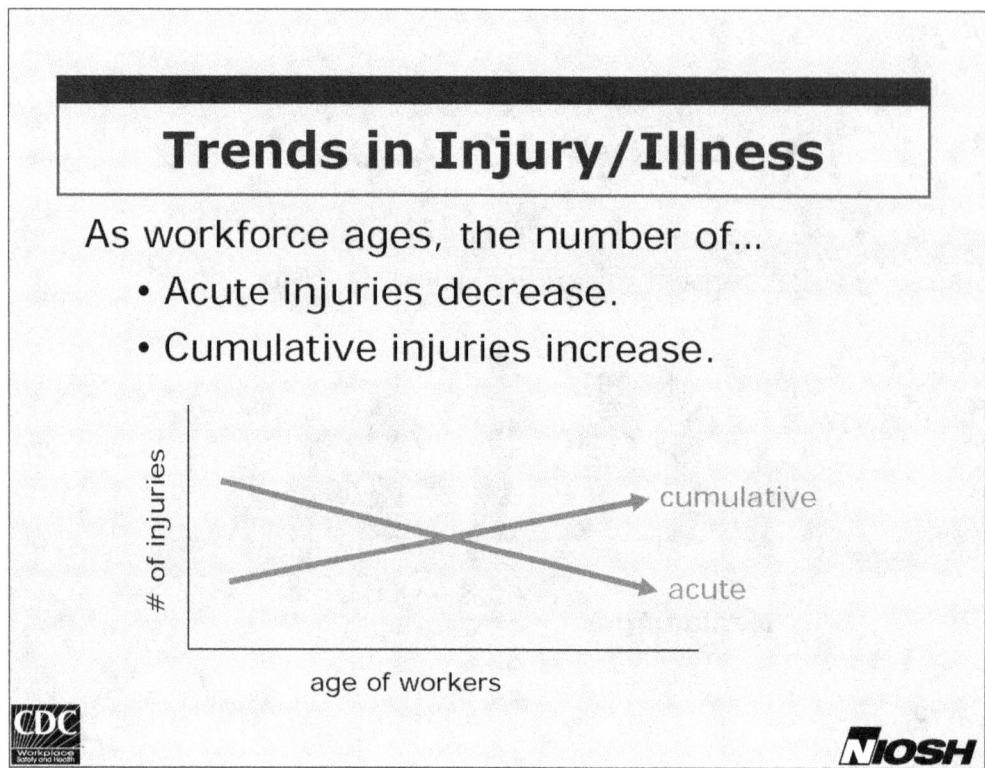

Key Points:

- Acute injuries (those that happen instantly, such as breaking an arm) are less common with age.

- Cumulative/musculoskeletal injuries (those that develop over time, such as tendonitis or carpal tunnel syndrome) are more common with age. They:
    - Can affect the muscles, tendons, ligaments, and nerves.
    - Are different from acute injuries because they develop slowly.
    - Can require more time for recovery as a worker ages.
    - Are difficult to identify early on.

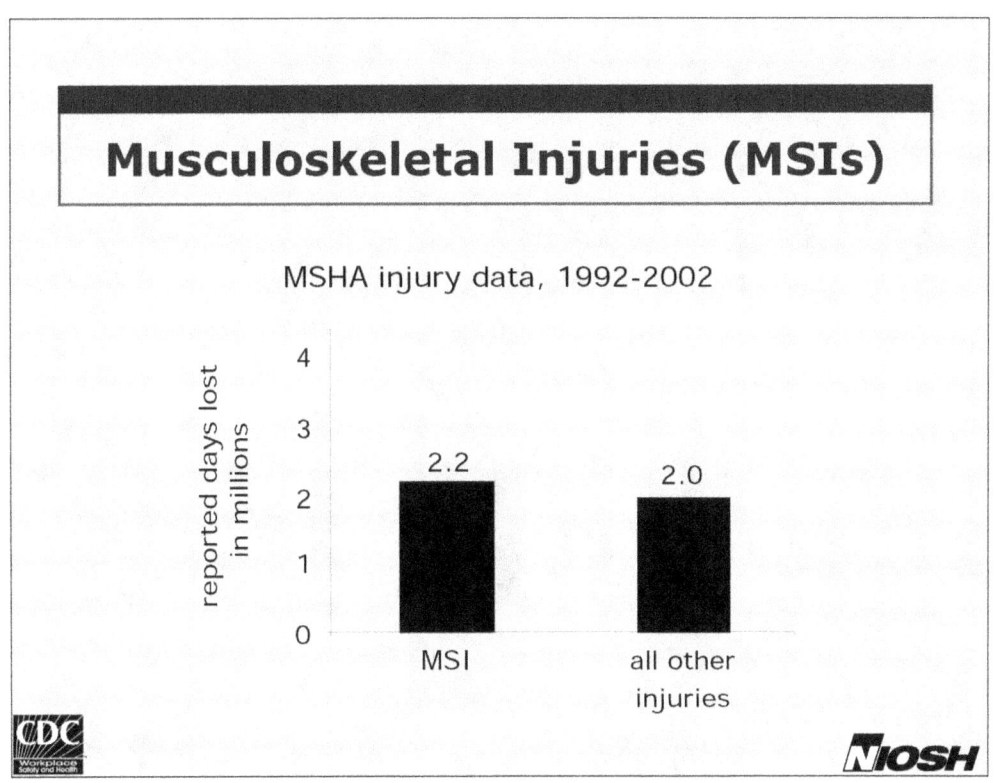

Key Points:

- Over the 10-year period of 1992–2002, over half of all workdays lost due to injury were musculoskeletal-related.
- On average, about 40% of all injuries that occurred during that period were musculoskeletal in nature.
- Because of the prevalence of musculoskeletal injuries (MSIs), there are several sections in this training that address MSIs.

NOTE: To assist companies who may want to start an ergonomics process, NIOSH has developed a training package called "Ergonomics and Risk Factor Awareness Training for Miners" (Information Circular (IC) 9497, NIOSH publication No. 2008-111). The overall objective of this training is to help reduce injuries and illnesses resulting from exposures to risk factors by increasing awareness of risk factors and encouraging miners to take action to report and reduce their exposures to risk factors. This training package was designed specifically for the mining industry and includes two components: Ergonomics and Risk Factor Awareness Training for Instructors; and a Guide to Conducting Ergonomics and Risk Factor Awareness Training. This training package, available in hard copy or electronic format, can be obtained by contacting 1-800-CDC-INFO or by downloading from the NIOSH Mining Web site (www.cdc.gov/niosh/mining).

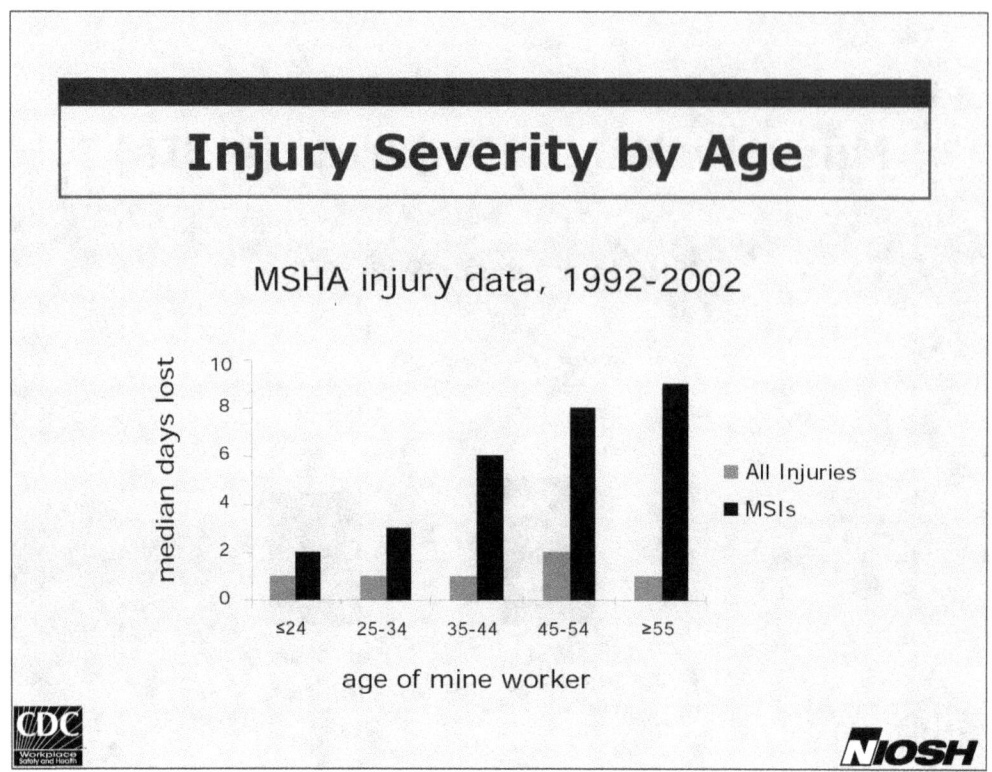

Key Points:

- The chart shows the median days lost per incident for all injuries and for MSIs broken down by age group.
- Days lost due to MSIs clearly increase with age.
- This supports the idea that workers require more time off for MSIs than for other types of injuries.
- 45- to 54-year-old miners with MSIs are off work four times longer than miners in their early twenties.

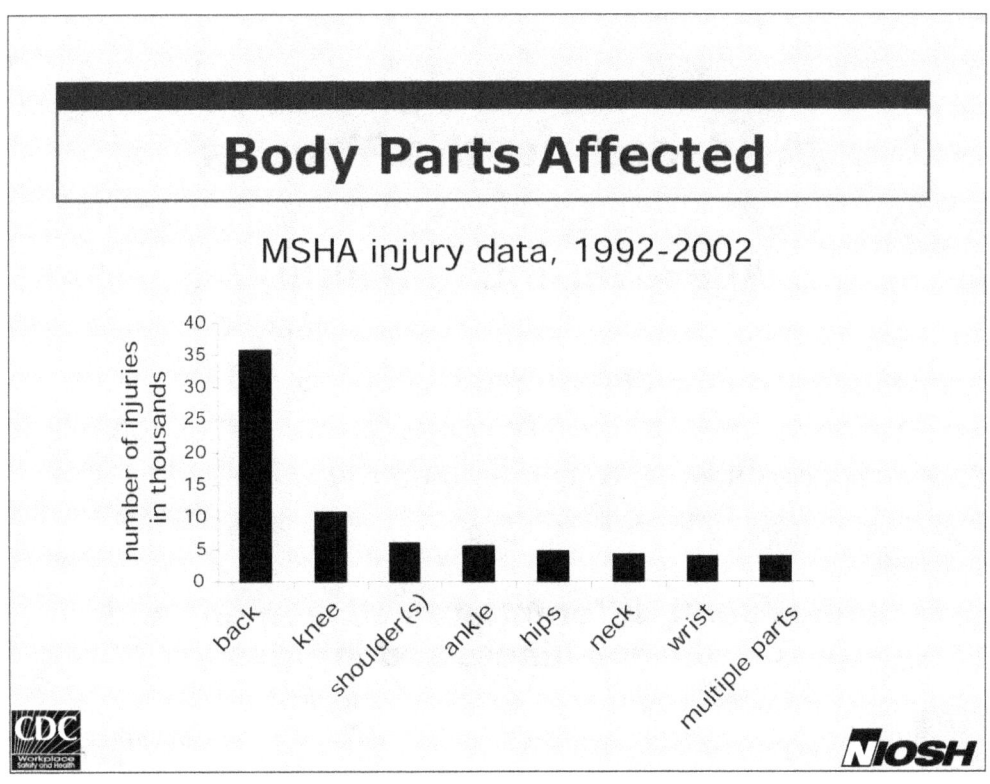

Key Points:

- The chart shows the parts of the body that are most commonly affected by MSIs.
- Back and knee injuries are the most common injuries reported.

Addressing the Truths

Age Awareness Training

- Informs about age-related changes
- Shows ways to modify worksites, tasks, or jobs
- Suggests ways to remain healthy

Bodies don't differentiate between on- and off-the-job illness and injuries.

Key Points:

The intent of this training is to:

- Inform workers about normal age-related changes.
- Instruct workers on how they can reduce the effects of these changes by modifying their worksites.
- Provide information to the workers about how to remain healthy as they get older.

- If you have an older workforce, you may have already seen the impact that years of working have had on some employees. Perhaps you are seeing a rise in sprains and strains or in slips, trips, and falls.
- If you have a younger workforce, you may be seeing employees with limited experience on the job. Now is the time to teach them to protect themselves and develop good habits as they work so they will remain safe and healthy for many years to come.

- These changes can be effective not only at the worksite, but also at home.

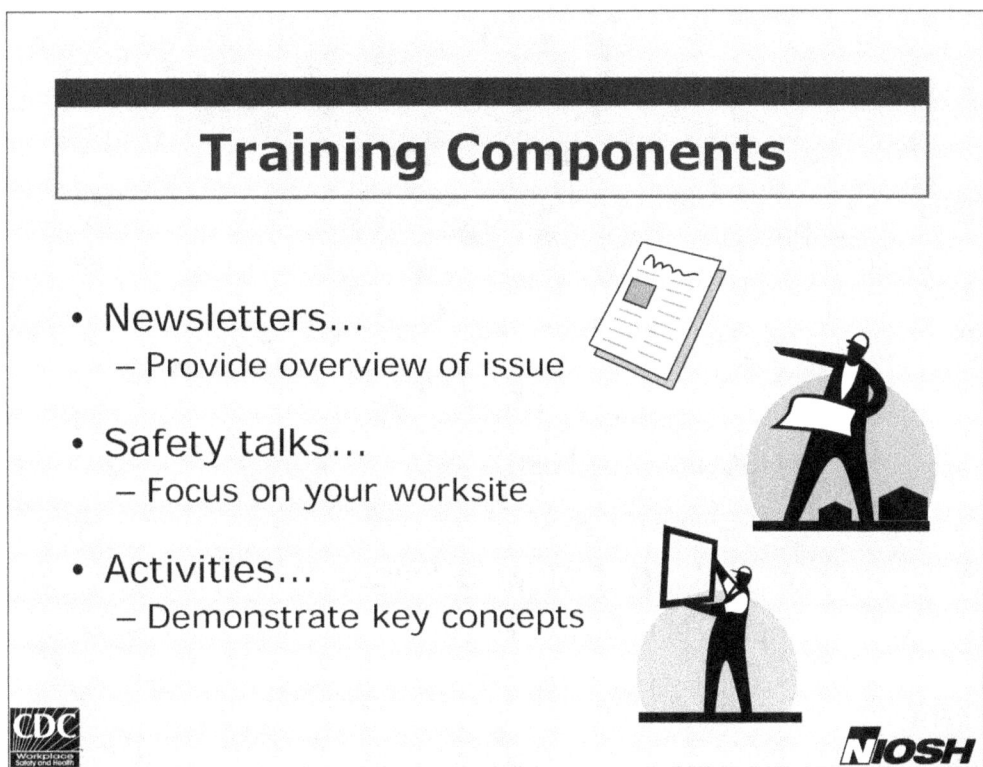

Key Points:

- The training consists of:
    - Newsletters that are two pages long and are meant to provide background information. Explain how the newsletters will be given to the employees at your site. They can be mailed to their homes or distributed in a company newsletter.

    - The safety talk guides are meant to be used during 10- to 15-minute sessions. They summarize information from the newsletter in a simpler, bulleted format and provide questions to help the team leader/supervisor/safety manager guide a discussion on the topic.

    - Every topic also has at least one activity. Most activities can be done in the field. The activities are intended to clarify one of the main points from the safety talks. Activity sheets explain how to prepare for and conduct the activity and its purpose.

Topics Included

- Introduction
- Vision
- Hearing
- Attention and Memory
- Musculoskeletal System
- Lower Back
- Work Capacity and Endurance
- Slips, Trips, and Falls
- Identifying High-Risk Tasks

Key Points:

- These are the topics that are addressed in the AAT.
- Some of the normal age-related changes can be exacerbated by environmental exposures (at home and at work).
- Engineering and administrative changes should be implemented to match task demands with changing worker capabilities.
- Improving the work environment will benefit all employees.
- Healthy aging applies to all employees.

Activity Instructions:

- The next group of slides covers these topics in more detail. For further information on each topic, please review the corresponding module.

Vision

Fast facts
- Most people experience vision changes with age.
- Most people over 40 can't bring close objects into focus.
- With aging, adapting to the dark becomes more difficult and peripheral vision decreases.

Workplace, task, or job modifications
- Provide bifocal safety glasses.
- Make signs large and with contrasting colors.
- Install lights in dark walkways and other transition areas.

Key Points:

- Most people experience some normal age-related vision change in their lifetime.
- For most people, it means getting reading glasses and having difficulty seeing in low-lit areas.

- To reduce the risk of vision-related injuries:
    - Be sure appropriate safety glasses are being used.
    - Consider vision when selecting signs and lighting.

Slide Animations:

- At first, only the "Fast facts" section will appear.
- Click for "Workplace modifications" to appear (the "Fast facts" section will turn gray).

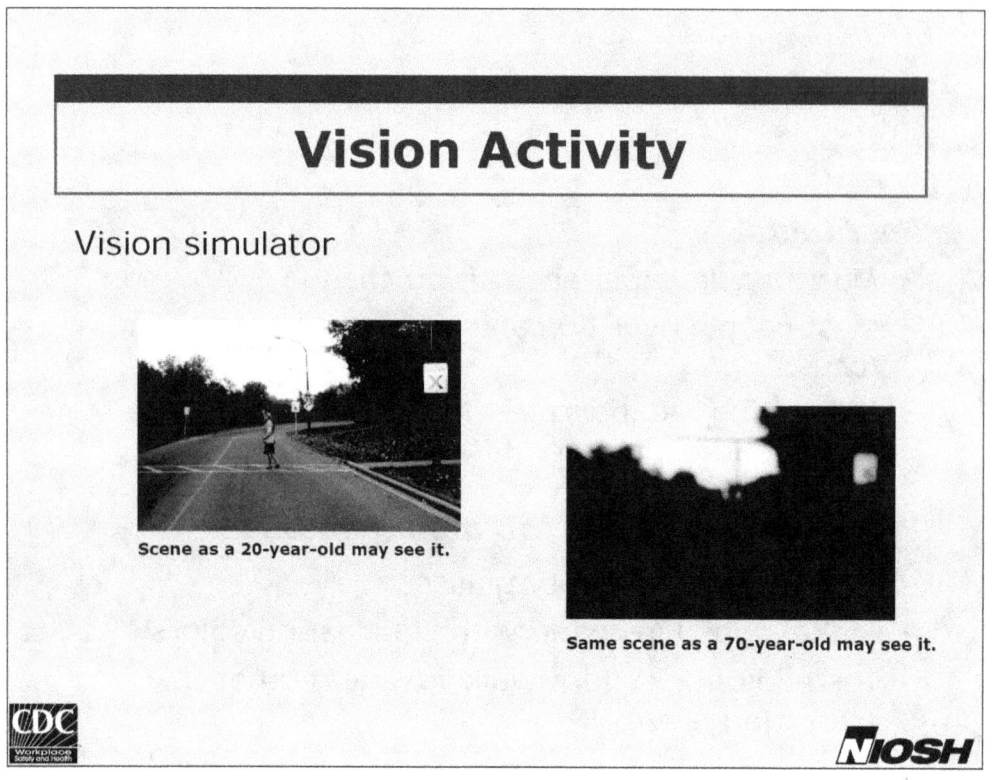

Key Points:

- As the eye ages, it goes through structural changes. This is why our eyesight does not remain as sharp as we age.
- 70% of people over age 45 wear glasses or contact lenses.

Activity Instructions:

- Have the audience look at each of the pictures and note the differences between the two.
    - The left photo shows what a young observer (20 years old) would see.
    - The right shows the same scene, but how an older observer (70 years old) would see it.

Hearing

Fast facts
- As people age, their ability to hear diminishes.
- Noise-induced hearing loss increases the problem and is irreversible.
- Men often lose ability to hear high-frequency sounds (above 4000 hertz).

Workplace, task, or job modifications
- Wear hearing protection in high noise areas.
- Use multiple types of warning alarms to accommodate workers with hearing loss.

Key Points:

- While hearing loss is the most prevalent occupational illness for miners, it is also highly preventable, especially if precautions are taken at an early age.
- Noise-induced hearing loss increases the problem and is irreversible.

- Noise is also an indirect safety hazard because it can "mask" important sounds like backup alarms and spoken warnings.

- To reduce the risk of a hearing-related injury:
    - Employees must follow the regulations with regard to noise exposure and hearing protection.
    - Multiple types of warning alarms should be used; this is called redundant coding. For example, an alarm might beep and a light might flash.

Slide Animations:

- At first, only the "Fast facts" section will appear.
- Click for "Workplace modifications" to appear (the "Fast facts" section will turn gray).

Hearing Activity

Self-Assessment

- Do you have to turn the volume up on the television?
- Do you frequently have to ask others to repeat?
- Do you have difficulty understanding when in groups or in noisy situations?
- Do you have to sit up front in meetings or in church in order to understand?
- Do you have difficulty understanding women or young children?
- Do you have trouble knowing where sounds are coming from?
- Are you unable to understand when someone talks to you from another room?
- Have others told you that you don't seem to hear them?
- Do you avoid family meetings or social situations because you "can't understand"?
- Do you have ringing or other noises (tinnitus) in your ears?

○ 3 or less = no symptoms of hearing loss
○ 3 to 5 = signs of slight hearing loss
○ 5 to 7 = signs of a moderate hearing loss
○ more than 7 = signs of significant hearing loss

Key Points:

- This activity allows people to do a quick check of their hearing. Any symptoms of hearing loss should be followed up with an audiologist.

Slide Animations:

- At first, only the "Self-Assessment" section will appear.
- Click for the scoring box to appear.

Activity Instructions:

- First, have the audience members read the statements and keep a count of their "yes" answers.
- Advance the slide (such that the scoring box appears), and ask the audience members to determine which category they fall into.

Attention and Memory

Fast facts
- Working memory begins to decline in early adulthood, and the rate of decline stays fairly constant throughout the lifespan.
- Decreases occur in tasks requiring divided attention or with distractions.

Workplace, task, or job modifications
- Provide cognitive aids like checklists.
- Eliminate divided attention tasks.

Key Points:

- Decreases in working (short-term) memory begin early in adulthood and continue through the later decades of life. There does not seem to be any evidence for an acceleration of the decline in the later decades [Hedden and Gabrieli 2004]. The decline in working memory can be compensated for (somewhat) by improved long-term memory. However, reducing the amount of information workers need to stay safe is always a good idea.
- Divided attention requires us to attend to two or more events simultaneously (e.g., monitoring the mine roof while operating a continuous miner, and monitoring the shuttle car, miner cable, methane levels, etc.).
- Whether or not these changes affect work performance depends on the extent of the individual's decreases in attention and memory and the difficulty of the work.

- To reduce the risk of an attention- and memory-related injury:
  - Review jobs to see where cognitive aids could be used (e.g., checklists for nonroutine tasks), or ascertain ways to eliminate divided attention tasks.

Slide Animations:

- At first, only the "Fast facts" section will appear.
- Click for "Workplace modifications" to appear (the "Fast facts" section will turn gray).

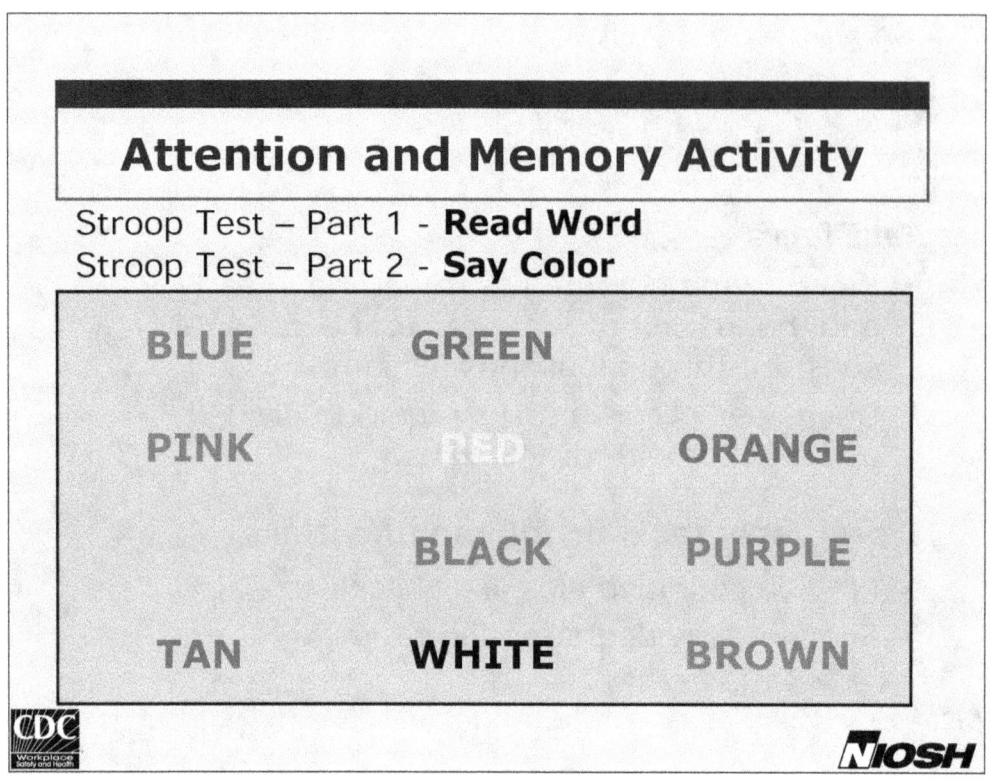

Key Points:

- The Stroop test demonstrates how performance decreases when we try to respond contrary to our practiced techniques.
- The test emphasizes how people can have difficulty inhibiting previously known information or preferences.

Slide Animations:

- First, only the top statement "Stroop Test – Part 1 - Read Word" will appear with the chart.
- Click the space bar and the second statement, "Stroop Test – Part 2 - Say Color," will appear.

Activity Instructions:

- You will need a stopwatch or a watch with time in seconds.
- Ask a volunteer to read the actual words presented in the chart (record the time it took him/her to read the words from start to finish).
- Advance the slide and ask him/her to say the color the word is written in (again record the time it took to read the color from start to finish).
- When comparing these two times, you should find that the time to read the written words is much shorter than the time needed to say the color in which each word is printed.

Musculoskeletal System

Fast facts
- Amount of muscle starts decreasing around age 30 and continues.
- Losses in muscle mass and strength can be improved by regular weight-bearing exercise.

Workplace, task, or job modifications
- Avoid bending and twisting, especially when lifting.
- Organize work area to avoid lifting or working above shoulders or below the knees.

Key Points:

- The musculoskeletal system is the body's system of bones, muscles, tendons, and ligaments.
- Injuries to this system are commonly referred to as "sprains and strains."
- Even though our aging bodies undergo a natural change where the amount of muscle tends to decrease, physical fitness helps reduce that decrease.

- To reduce the risk of an MSI injury:
  - Find comfortable neutral postures for work tasks.
  - Avoid bending and twisting.
  - Reduce tasks that require lifting or working above shoulders or below the knees.

Slide Animations:

- At first, only the "Fast facts" section will appear.
- Click for "Workplace modifications" to appear (the "Fast facts" section will turn gray).

Lower Back

Fast facts
- Part of the body most often affected by MSIs
- Back injuries have high recurrence rates (40%-70%).
- Back injuries can result from a variety of causes.

Workplace, task, or job modifications
- Keep object being lifted close to the body.
- Add mechanical-assist devices (such as hoists or jacks) to reduce demands of heavier loads and awkward lifts.

Key Points:

- Back injuries are prevalent in mining.
- These injuries are painful and costly to the individual and costly to companies.

- To reduce the risk of a back injury:
    - Keep the object being lifted close to the body.
    - Don't lift heavy or awkward-sized objects alone.

Slide Animations:

- At first, only the "Fast facts" section will appear.
- Click for "Workplace modifications" to appear (the "Fast facts" section will turn gray).

Work Capacity and Endurance

Fast facts
- Starting in the mid-twenties, the amount of physical effort that workers can sustain without fatigue decreases.
- Reduced endurance results in more rapid fatigue when working at the same intensity.

Workplace, task, or job modifications
- Take rest breaks during demanding tasks.
- Schedule easier tasks in between those requiring heavy exertion.
- Move materials with mechanical aids or carts.

Key Points:

- Declines in work capacity and endurance generally start in the mid-twenties.
- This reduced endurance to physically demanding tasks will result in more rapid fatigue, which may make it more likely for workers to make mistakes or become injured.

- To reduce the risk of a work capacity-related injury:
    - Perform highly physically demanding jobs for shorter periods of time.
    - As you age, expect to take more time than you used to for the same task.

Slide Animations:

- At first, only the "Fast facts" section will appear.
- Click for "Workplace modifications" to appear (the "Fast facts" section will turn gray).

Slips, Trips, and Falls

Fast facts

- The number and severity of slips and falls increase as people age.
- Ground surface transition areas are the most likely place for a slip, trip, or fall injury.

Workplace, task, or job modifications

- Make repairs to holes and uneven pavement.
- Put warnings in areas with slip/trip hazards.
- Install adequate lighting in walkways.

Key Points:

- Older people experience more slips and falls and are more severely injured.
- Changes in balance and loss of lower-extremity strength and dexterity may cause these increases.
- Slips and falls can largely be prevented by ensuring that there are safe walking surfaces.

- To reduce the risk of a slip, trip, or fall injury:
    - Check walkways and aisle ways so that they are free from debris, excess water, and ice.

Slide Animations:

- At first, only the "Fast facts" section will appear.
- Click for "Workplace modifications" to appear (the "Fast facts" section will turn gray).

Identifying High-Risk Tasks

Trainees are given workplace scenarios that include hazards associated with normal age-related changes.

Trainees identify hazards that put an age-diverse workforce at risk.

Key Points:

- This exercise provides opportunities to evaluate risk of injury for a work task scenario.
- Trainees are given workplace scenarios that include hazards associated with normal age-related changes.
- Then the trainees identify potential risk factors that may be affected by age-related changes.
- The trainees should suggest changes that could reduce each risk of injury.
- This activity moves trainees to the next level by providing practice in assessing worksites and tasks where multiple potential hazards exist.

Implementing Training

- Distribute newsletters.
- Give safety talks and activities.
- Review completed safety talk guides and evaluation forms.
- Discuss the issues with your crew and coworkers.

Key Points:

- The next steps to implement the training are:
    - Schedule newsletter distribution and safety talks with related activities.
    - Review completed safety talk guides and evaluation forms.
    - Discuss the age-related changes and related risks with your crew and coworkers.
- Review the AAT Instructor's Guide for details on this process.

For More Information

William L. Porter
412-386-5222
WLPorter@cdc.gov

The findings and conclusions in this presentation have not been formally disseminated by the National Institute for Occupational Safety and Health and should not be construed to represent any agency determination or policy.

NOTE: You may want to change the contact information to the champion for your location.

Topic:
Introduction

Age Awareness News

Who Are the Baby Boomers?

Articles about the country's aging workers are prevalent in magazines and newspapers. A lot of attention has been paid to the abilities, spending habits, and health of a group of people known as the Baby Boomers. Many miners fit into this age group of approximately 80 million people born in the late 1940s, 1950s, and early 1960s.

The mining workforce parallels national demographics. The pie chart shows that more than half of mine workers in 2005 were over 44 years old (Bureau of Labor Statistics, Current Population Survey). Many of these miners have worked for over 20 years in mining and have accumulated the expertise as well as possible cumulative health effects that come with such a career.

This newsletter is the first in a series about age-related safety and health topics. Each discusses normal age-related changes that occur over a lifetime. Learning how to mediate physiological changes through workplace modifications and improved health is the purpose of this training.

Percentage of Mine Workers by Age (2005)

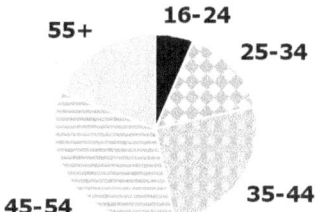

Key Points:

- No matter how old or young you are, you can benefit from understanding normal age-related changes.

- The mining workforce is aging.

- Aging affects physical and mental abilities.

- An older workforce's strengths, such as experience and corporate knowledge, are extremely valuable.

How It Works

As a workforce ages, it improves in some ways and is more at risk in others. So what's young and what's old?

In the United States, workers must be at least 19 to hold certain jobs as defined by the Fair Labor Standards Act. That is also a common age for high school graduation and first full-time employment.

At the other end of the spectrum, "older worker" can be defined many ways. Often workers are eligible to retire when they are in their sixties, but employees are covered by the Age Discrimination in Employment Act when they reach 40 years old.

There is no standard definition of older or younger worker. When considering age-related workplace issues, the number isn't important.

Normal age-related changes occur at different times for different people. While an older workforce will have different risks and strengths than a younger one, those risks and strengths will vary across individuals.

In a demanding workplace, the healthiest people can remain working the longest. Older miners are likely to be physically younger than their age suggests. On the other hand, workers who have been stressing their bodies over many years may find the cumulative effect leads to aches, pains, or more serious injuries.

So when thinking about when you became or will become an "older worker," consider both your actual age and what you have done and will do to lessen age-related risks.

A younger construction worker teased an older worker about his age. Older worker: "Why don't you put your money where your mouth is? I'll bet a week's wages that I can haul something in a wheelbarrow over to that building and you won't be able to wheel it back." Braggart: "You're on, old man. Let's see what you got." The old man grabbed a wheelbarrow by the handles and said, "All right. Get in."

—Michael Brickey, Ph.D.

Normal Age Changes

Some age-related changes are commonly known. That's why people make jokes about needing longer arms to read the paper as their vision changes.

Others, such as changes in the ability to process new information, may be less well-known and understood.

Future newsletter issues will focus on the types of changes that occur as the body ages. These topics include:

- Vision
- Hearing
- Attention and Memory
- Musculoskeletal System
- The Lower Back
- Work Capacity
- Slips and Falls

While some diseases do occur more frequently in older people, this training is about normal changes that everyone goes through to some degree at some age and not about disease-related changes. The age when any specific change happens can vary among individuals.

Being aware of age-related changes will allow you to take advantage of positive changes and to postpone or alleviate negative ones.

Changing the Workplace to Reduce Risk

Age-related changes can affect your risk for injury from certain job tasks. Some age-related changes may make it difficult for a specific task to be done in a certain way. Also, some jobs may speed the onset of certain normal age-related changes.

In most cases, changes can be made to job tasks, methods, and/or jobsites to reduce both of these types of risks.

For example, age-related vision changes can make detailed work more difficult. However, a simple change like adding more lighting can allow a worker to overcome this vision limitation and continue to do the job.

Hearing loss is another good example. Age-related hearing loss is normal, but workers overexposed to high levels of noise will have a significantly greater problem than those who are not overexposed. Reducing the noise at a worksite will help workers maintain hearing at as high a level as possible as they age.

As you read future newsletters, assess your workplace for locations and tasks that may be associated with age-related changes. If needed, modify them to protect yourself and others as all of you age.

Preventive Maintenance for People

"If I had known I was going to live this long, I'd have taken better care of myself when I was young." —Anon.

How often do we think about what we are doing to protect our health? No matter how old you are, you can take action to help you age healthfully.

Preventive maintenance on equipment reduces later, more costly problems and keeps machinery running smoothly. Preventive maintenance on people can do the same.

As we age, we all want to be able to continue doing the things we enjoy, like throwing a football, traveling, or walking through the woods. Preventive actions now will help you enjoy those activities later in life.

It's up to you to use what you learn from this program to invest in your future.

Topic: Introduction

Topic:	
Introduction	# Age Awareness Safety Talk Guide

Introduction

- The relationship between aging and work is important to employees of all ages.
- This is the first in a series of safety talks about this topic.
- During each talk, we will discuss how a topic covered in the newsletter is important at our worksite.
- Future talks will cover specific age-related changes to:
    - Vision
    - Hearing
    - Attention and Memory
    - Musculoskeletal System
    - The Lower Back
    - Work Capacity
    - Slips and Falls
- We will assess our jobsite and tasks to see why age-related changes might matter to us.
- Keeping our employees healthy throughout their work lives and into their retirement years is our vision.

The purpose of this training is to discuss normal age-related changes and to demonstrate how those changes can be mediated through reducing "job risks" and "life risks."

Assessing Our Age-related Concerns

- As we go through the topics during these talks, we will find out at what age physical changes typically take place, understanding that every individual is different.
- Each of us will have to determine which, if any, of the changes we are experiencing and to what degree.
- Awareness will allow us to use our strengths and to modify our situations to compensate for changes.

Ways to Reduce Risk

- During this training, we will learn to look at our jobs to see if better methods exist to reduce the risk of injury.
- When possible, we'll make needed changes. When we can't, we'll make management aware of the need for change.
- Management has shown a willingness to listen to these kinds of issues by bringing this program to our worksite.
- The best situation involves modifying potentially risky situations before they become a problem.

What ages are workers at this workplace? ... in this work group or crew?
(For example: many older workers, a lot of young/inexperienced workers, a mix)

_____

_____

_____

_____

_____

What safety and health issues might be related to our ages?
(For example: inexperience leading to unsafe acts, increase in back injuries with older employees)

_____

_____

_____

_____

_____

What can we do to age healthfully?
(For example: exercise regularly; specific examples will be given in later safety talks)

_____

_____

Topic: Introduction

| Topic: Introduction | # Age Awareness Activity |

Age Awareness Status Check

This first age awareness activity introduces concerns that will be covered by this training program. It can also provide a baseline for the trainee's current thinking and knowledge about age-related changes and how they impact and are impacted by work.

Equipment Needed

- Copies of the Introduction Activity Questions
- Pencils or pens

Procedure

1. Explain that this activity is not a test. It is a way to assess what everyone currently thinks about age and work issues.
2. After trainees have completed the form, discuss their answers. Tell them they will learn more about these topics during this training program.
3. Include the answers to the last question in your review process for potential workplace and task changes.
4. You can collect the forms and repeat this activity after training to determine if and how opinions and knowledge have changed.

Why Is This Important?

Everyone learns to think about getting older based on the beliefs of their culture and their experiences with family, coworkers, the media, schools, and other social settings. Without consciously thinking about their beliefs, employees won't know whether or not what they believe about aging and work matches the reality of their workplace. This activity was designed to get employees thinking and to provide a place for trainers to start discussions about aging issues.

For More Information

Stein D, Rocco TS [2001]. The older worker. Myths and Realities No. 18. Columbus, OH: Ohio State University, Center on Education and Training for Employment. Available at: http://www.cete.org/acve/docs/mr00033.pdf

Aging Myths and Emerging Realities. Pittsburgh, PA: University of Pittsburgh, Institute on Aging. http://www.aging.pitt.edu/family-caregivers/myths/default.asp

Your job title: _____

Age _____   Age when you started in mining? _____

At what age does a miner become: Middle-aged? _____   Old? _____

At what age do miners usually retire? _____

Please circle the number that most closely matches your opinion on each statement below.

	Strongly Agree	Agree	Disagree	Strongly Disagree
Some jobs are better suited to younger miners.	4	3	2	1
Some jobs are better suited to older miners.	4	3	2	1
Age-related changes impact how people work.	4	3	2	1
Job performance declines with age.	4	3	2	1
Younger miners get hurt more often.	4	3	2	1
It's hard to teach new skills to an older miner.	4	3	2	1
Younger miners are a better training investment.	4	3	2	1
Jobs can be changed to make them better fit the capabilities of older miners.	4	3	2	1
The following are important to doing **my** job:				
♦ Vision	4	3	2	1
♦ Hearing	4	3	2	1
♦ Muscle strength	4	3	2	1
♦ Dexterity	4	3	2	1
♦ Stamina	4	3	2	1
♦ Physical flexibility	4	3	2	1
♦ Memory	4	3	2	1
♦ Back health	4	3	2	1
Certain aches and pains are just part of my job.	4	3	2	1
My job can be changed to reduce aches and pains.	4	3	2	1
My job is difficult for a younger worker.	4	3	2	1
My job is difficult for an older worker.	4	3	2	1

Can your job be changed to better fit an older workforce?   Yes   No
   If yes, please explain how it could be changed on the back of this page.

Topic: Introduction

Topic:
Vision

Age Awareness News

What Do You See?

People rely heavily on vision to carry out their work. Protecting your eyes is important because your eyes protect you!

Most people will go through some normal age-related vision changes during their lifetime. These changes may diminish their ability to see. If your vision is impaired, you may find yourself in dangerous situations.

Two normal changes that occur are decreased abilities to discriminate between objects in dimly lit situations and to see objects at close range. Depth perception also decreases with age. These changes can affect your ability to safely operate equipment or perform maintenance activities.

There are also some eye conditions, such as cataracts or glaucoma, that are more common in older people and may require medical attention.

Key Points:

- As people age:
  - The ability to distinguish details at short distances decreases.
  - The accuracy of speed perception decreases.
  - Resistance to glare declines.
  - Depth perception decreases.
- 70% of people over age 45 need to wear glasses.
- Most people find that sometime between the ages of 40 and 50 close objects can no longer be brought into focus without corrective lenses.

How It Works

When you look at something, light rays are reflected from the object. The light rays are bent, refracted, and focused by the cornea, lens, and vitreous gel in the eye.

The lens of the eye makes sure the rays come to a sharp focus on the retina. The resulting image on the retina is upside down.

At the retina, the light rays are converted to electrical impulses that are transmitted through the optic nerve to the brain. There the image is perceived in an upright position.

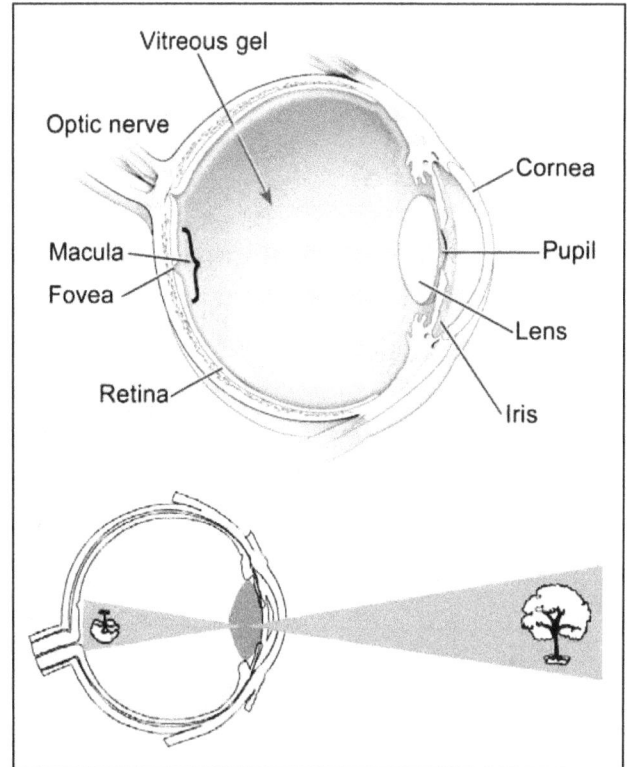

A machinist (age 52) was about to make a critical cut on a small steel post spinning on the lathe in a tool shop. He wanted this part to be perfect, so he took off his safety glasses and put on his drugstore "cheaters" so he could better see the exact point to make the cut.

The moment he touched the steel with the cutter, the steel kicked back toward his face and imbedded into his eye.

In a split second, without the protection of his safety glasses, he was blinded in one eye.

[La Haye and Sustello 2001]

Topic: Vision

Normal Age Changes

Age-related changes occur in the iris, the lens, the vitreous gel, and the retina's photoreceptor cells.

As the muscles in the iris weaken, the pupil becomes smaller and slower to change. The smaller pupil allows less light to enter the eye. By age 60, the amount of light reaching the photoreceptors is 33% of that at age 20.

The lens yellows and hardens, causing a number of problems such as decreased visual acuity and difficulty seeing blues. Between the age of 40 and 50, the inelasticity of the lens hinders the ability to focus on close objects (presbyopia). By 60, the ability to focus on objects within 3 feet disappears.

After age 40, changes in the lens and the vitreous gel cause the resistance to glare to decline 50% every 12 years. Other conditions that can arise include: problems seeing when going from light to dark areas, a decrease in peripheral vision, and decreased ability to accurately judge the speed of moving objects.

Decreased depth perception can lead to a problem called "looming," which makes determining the distance between two objects difficult.

Changing the Workplace to Reduce Risk

The following tips provide some workplace improvements to accommodate age-related vision changes:

Improve signs by using strong contrasts (black against white or white against green).

Encourage people to take their time when transitioning between dark and light areas, walking over uneven ground, or climbing steps.

Design important tasks near the direct line of sight to avoid dependence on peripheral vision.

Reduce glare by using more than one light source and avoiding glossy surfaces.

Use daytime running lights to enhance motion perception and speed estimation.

Increase lighting in dimly lit areas.

Promote using safety glasses with bifocals and/or lens with UV protection, as needed.

Preventive Maintenance for People

One of the most important things you can do to protect your eyesight is to get routine eye exams to determine your need for corrective lenses, address possible injuries, and identify diseases.

The best way to protect your eyesight at home or on the jobsite is with protective eyewear. It is important to match the method of eye protection with the hazards involved with the task.

- When appropriate, wear UV-filtering sunglasses to protect your eyes from harmful sun rays.
- Wear certified safety glasses whenever your eyes are at risk of being injured from debris or flying objects.
- Use goggles or full face shields when working with chemicals to protect from possible exposures.
- Welding shields should be used to protect the eyes from harmful welding rays.

| Topic: Vision | # Age Awareness Safety Talk Guide |

Introduction

- Most people go through some normal age-related vision change in their lifetime.
- 70% of people over age 45 need to wear glasses.
- For many people over 40, close objects can't be brought into focus without corrective lens (often bifocals are used).
- Other age-related vision changes include:
  - Problems seeing when going from light to dark areas
  - Loss of peripheral vision
  - Diminished ability to accurately judge the speed of objects and to judge distances between objects
- Some vision changes can increase the risk of workplace injury.
- Taking vision changes into account when designing jobs can make jobs safer and more efficient.

The need for bifocals carries over into the workplace.

Never remove safety glasses to put on "cheaters."

ANSI Z87.1-compliant safety frames can be fitted with impact-rated Plano lenses with molded-in magnifying bifocals.

Some Tasks Affected by Vision Changes

- Walking through areas where the lighting changes abruptly
- Using power tools that require focusing on close objects
- Driving when judgment of speed or distance is needed

Ways to Reduce Risk

- Lessen eyestrain by increasing task lighting in low-light areas.
- Reduce glare by rearranging work areas or redirecting light sources.
- Use large lettering and contrasting colors for text and background on warning signs.
- Be aware that some people have trouble judging distances.
- Wear corrective lens as prescribed.
- Wear safety glasses with bifocal lens or ultraviolet (UV) protection, as needed.

What are some tasks with vision-related risks on this worksite?

(For example: abrupt lighting changes, bright sunlight, tasks that require depth perception)

_____

_____

_____

_____

_____

How can this worksite be changed to reduce vision-related risk?

(For example: wear safety glasses, increase lighting, increase size of print in instructional materials)

_____

_____

_____

_____

_____

What can we do to reduce risk to our vision?

(For example: get routine eye exams, increase lighting at home)

_____

_____

Topic: Vision

Topic:	
Vision	# Age Awareness Activity

With age:

- The lens scatters light across the retina and can cause increased sensitivity to glare.

- The lens hardens and yellows.

- The amount of light reaching the retina decreases.

Structural Changes of the Eye

As the eye ages, it goes through structural changes. It is for this reason that our eyesight does not remain as sharp as we age.
In fact, 70% of people over age 45 wear glasses or contact lenses.

Equipment Needed

- Two images (provided on the next page)
- Color printer or color computer monitor to display images

Setup Instructions

Print out or display the two images found on the next page in color.

Procedure

Take a look at each of the pictures provided, and note the differences between the two. One picture shows what a young observer (20 years old) would see, and the other shows what the same scene might look like to an older observer (70 years old). Note the decline in brightness and clarity of the image.

Why Is This Important?

Changes in vision like this are common in the normal aging process, but it is still important to be aware and attempt to accommodate for them. Eyesight is perhaps the sense that humans rely on the most. Therefore, changes in eyesight as we age need to be taken seriously and accommodated for in the workplace. A couple of things that should be considered for aging workers are the level of workplace lighting and print size of signs and text in the work area.

For More Information

NIOSH Safety and Health Topic – Eye Safety:
http://www.cdc.gov/niosh/topics/eye/

Nolan DE [2002]. Normal age-related vision loss and related services for the elderly. Available at: http://hubel.sfasu.edu/research/donia/aging_vision_front.htm

Activity Pictures

These pictures show a woman walking across the road at dusk.

(Used with permission from the University of Calgary Vision and Aging Laboratory, Calgary, Alberta, Canada.)

Scene as a 20-year-old may see it.

Same scene as a 70-year-old may see it.

Topic: Vision

Topic: Vision

Age Awareness Resources

Additional Resources

Medline Plus – Vision Impairment and Blindness:
http://www.nlm.nih.gov/medlineplus/visionimpairmentandblindness.html

National Eye Institute: http://www.nei.nih.gov

NIOSH Safety and Health Topic – Eye Safety: http://www.cdc.gov/niosh/topics/eye/

Relevant Federal Regulations (intended as a guide, not a comprehensive list)

Title 30 CFR: Mine Safety and Health Administration (MSHA)

56.15004 Eye protection: http://www.msha.gov/30cfr/56.15004.htm

56.20011 Barricades and warning signs: http://www.msha.gov/30cfr/56.20011.htm

19.6 Specific requirements for approval of lamps: http://www.msha.gov/30cfr/19.6.htm

75.1719-1 Illumination in working places: http://www.msha.gov/30cfr/75.1719%2D1.htm

56.17001 Illumination of surface working areas: http://www.msha.gov/30cfr/56.17001.htm

Title 29 CFR: Occupational Safety and Health Administration (OSHA)

1910.133 Eye and face protection:
http://www.osha.gov/pls/oshaweb/owadisp.show_document?p_table=STANDARDS&p_id=9778

1926.102 Eye and face protection:
http://www.osha.gov/pls/oshaweb/owadisp.show_document?p_table=STANDARDS&p_id=10665

1910 Subpart I, Appendix B Personal protective equipment:
http://www.osha.gov/pls/oshaweb/owadisp.show_document?p_table=STANDARDS&p_id=10120

1910.37 Maintenance, safeguards, and operational features for exit routes:
http://www.osha.gov/pls/oshaweb/owadisp.show_document?p_table=STANDARDS&p_id=9725

1917.17 Railroad facilities:
http://www.osha.gov/pls/oshaweb/owadisp.show_document?p_table=STANDARDS&p_id=10359

Topic: Hearing

Age Awareness News

Do You Hear What I Hear?

By age 65, 50% of miners have enough hearing loss to cause difficulty with speech communication.

Many people lose their hearing at a consistent rate starting at age 60. Often men lose the ability to hear higher-frequency sounds rather than lower-frequency sounds. These losses can be accelerated by exposure to noise at home or at work.

Hearing loss can have a variety of causes, such as genetics, injury, illness, and the natural aging process.

Noise exposure is the most important factor in hearing loss for miners. Losses are also linked to factors such as cardiovascular disease, smoking, medications, and diet.

Loss of hearing can isolate people and cause communication and safety concerns at work and at home.

Any noticeable change in hearing should be evaluated by a medical professional immediately.

How It Works

The outer ear (pinna) collects sound waves and sends them through the ear canal to the eardrum.

Incoming sound waves cause the eardrum to vibrate, which sets the three tiny bones in the middle ear into motion.

The motion of these bones causes the fluid in the inner ear or cochlea to also vibrate.

Vibration of the inner ear fluid then causes the hair cells in the cochlea to move. The hair cells change this movement into electrical signals. These electrical impulses are transmitted by the hearing (auditory) nerve to the brain where they are interpreted as sound.

Disorders of the inner ear or auditory nerve are called sensorineural hearing loss.

Obstruction or disease in the outer or middle ear that affects the transmission of sound waves is called conductive hearing loss.

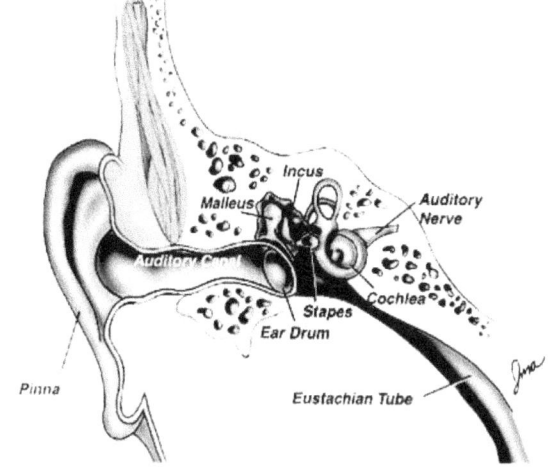

Source: National Institute on Deafness and Other Communication Disorders.

Key Points:

- Approximately 10% of all middle-aged adults suffer hearing loss at a magnitude that interferes with conversation.

- Since noise-induced hearing loss is preventable, it is important to wear hearing protection when needed.

- If you hear ringing in your ears after exposure to a loud noise, it means that damage is being done to your hearing.

Normal Age Changes

Most age-related hearing loss, also called presbycusis, is caused by damage to the hair cells in the cochlea.

People with this type of hearing loss will notice problems such as understanding conversations, especially when there is background noise, and hearing higher-frequency sounds such as female voices.

Normal loss starts to occur at the rate of 1 dB per year every year after age 60.

Hearing loss may occur slowly at first, so the person may not even be aware of it. Because of this, regular check-ups are key to detecting early stages of hearing loss.

Changing the Workplace to Reduce Risk

Make sure everyone is wearing hearing protection in posted high noise areas.

When possible, reduce background noise so that you can better understand conversations.

Investigate noise-dampening options on equipment. Consider better sound-insulating materials for cab interiors.

Encourage workers to take home their hearing protection for use when mowing the lawn or doing construction work.

Provide redundant signals, such as alarms and flashing lights to alert workers about equipment location.

Use and maintain existing engineering noise controls, such as cabs and mufflers.

To be better understood, speak at a reasonable and consistent rate (limit to two words per second).

When possible, adjust the volume of phones, computer speakers, etc.

Supplement auditory signals with information presented in another way. For instance, use vibration or flashing light signals.

Preventive Maintenance for People

One of the most important things you can do to protect your hearing is to wear the appropriate hearing protection both at work and at home.

- Follow all company policies regarding when and where hearing protection is needed.

- Encourage your family to protect everyone by keeping the car radio/stereo at a safe volume.

- Have your hearing assessed with routine annual exams.

- Participate in hearing conversation programs at work.

- Use hearing protection at work and at home when you are around loud noise.

- Protect yourself while operating power tools or while doing hobbies, such as shooting firearms or listening to music.

Hearing aid use must start with proper training.

Without proper training, 25%–50% of people fitted with hearing aids refuse to wear them because of lack of comfort or function.

Consultation with trained professionals will help ensure successful use of hearing aids.

Topic: Hearing

Topic: Hearing

Age Awareness Safety Talk Guide

Introduction

- Noise-induced hearing loss is preventable.

- Most people experience hearing loss throughout their lifetime. However, this varies considerably among the population, and many people experience very little hearing loss.

- Understanding speech becomes more difficult for older adults, especially when there is background noise.

- Men tend to lose the ability to hear high-frequency sounds (sounds above 4000 hertz).

- Older miners have much higher rates of hearing loss than would be predicted by natural age-related hearing loss.

- Noise exposure is the most important factor in hearing loss for miners.

Some Abilities Affected by Hearing Loss

- Hearing warning signals
- Understanding when someone is speaking
- Listening to equipment for cues to how well it's working

Ways to Reduce Risk

- Stay away from noise sources and use quieter products instead of relying solely on hearing protection to protect you.

- Minimize background noise. Simply making things louder doesn't always help because both the signal and the noise are increased.

- Wear appropriate and properly fitting hearing protection in any noisy area.

- Get hearing tests yearly.

Approximately 10% of all middle-aged adults have hearing loss that interferes with their conversations.

By age 65, the percentage increases to 50% of all men and 30% of all women.

Unprotected exposure to loud noises will cause those losses to occur at a younger age.

What are some tasks with hearing-related risks at this worksite?
(For example: tasks involving auditory warning signals, the need to understand speech, loud noise)

___

How can this worksite be changed to reduce hearing-related risk?
(For example: redundant signals, use of sound-dampening materials, purchase quieter equipment)

___

What can we do to reduce the risk of hearing loss?
(For example: hearing tests, wear ear protection during loud off-the-job hobbies, close cab doors)

___

Topic: Hearing

| Topic: Hearing | # Age Awareness Activity |

Noise-induced hearing loss is completely preventable.

Always follow company hearing conservation policies.

Wear hearing protection at home, too!

Hearing Self-Assessment

Age-related hearing loss (known as presbycusis) is common in people over 50. This activity allows people to do a quick check of their hearing. Any symptoms of hearing loss should be followed up with a licensed professional.

Equipment Needed

- Copies of the Hearing Self-Assessment
- Pencils or pens

Procedure

1. Explain that the purpose of the Hearing Self-Assessment is for trainees to think about whether or not they have experienced any symptoms of hearing loss.
2. Tell them the Self-Assessment forms will not be collected, so there will not be a record of their responses.
3. Encourage trainees to seek professional advice if they have any possibility of hearing loss.
4. Distribute the Self-Assessment forms.

Why Is This Important?

Hearing loss can happen so gradually that the person experiencing it doesn't notice the change. Taking a minute to think about hearing loss-related symptoms could lead someone to seek needed professional evaluation.

For More Information

Hearing Loss Association of America: http://www.hearingloss.org/

Hearing Self-Assessment

The following questions allow you to do a quick check of your hearing.

Check all that apply:

- Do you have to turn the volume up on the television?
- Do you frequently have to ask others to repeat?
- Do you have difficulty understanding when in groups or in noisy situations?
- Do you have to sit up front in meetings or in church in order to understand?
- Do you have difficulty understanding the voices of women or young children?
- Do you have trouble knowing where sounds are coming from?
- Are you unable to understand when someone talks to you from another room?
- Have others told you that you don't seem to hear them?
- Do you avoid family meetings or social situations because you "can't understand"?
- Do you have ringing or other noises (tinnitus) in your ears?

Scoring

If you checked:
- Fewer than 3 of the questions – no symptoms of hearing loss present
- Between 3 and 5 questions – you may have a slight hearing loss*
- Between 5 and 7 questions – you have a moderate hearing loss*
- More than 7 questions – you may have a significant hearing loss*

*To determine the exact degree of hearing loss, you should have your hearing evaluated by an audiologist.

Adapted from the Hearing Loss Association of America, Self-Assessment Hearing Test (http://www.hearingloss.org/learn/self-assessment.asp).

Topic: Hearing

Topic:	
Hearing	# Age Awareness Activity

NIOSH Hearing Loss Simulator

The NIOSH Hearing Loss Simulator is a Windows-based program that displays a "control panel" for playing sounds while adjusting simulated effects of noise and aging. A simulated individual's age can be entered along with years of exposure to noise (in A-weighted decibels). The effects are shown visually on the frequency band control panel and sound level display screen while the user listens to the audio playback.

Noise-induced hearing loss is completely preventable.

The NIOSH Hearing Loss Simulator allows you to experience different types of hearing loss.

You don't have to experience noise-induced hearing loss—wear hearing protection!

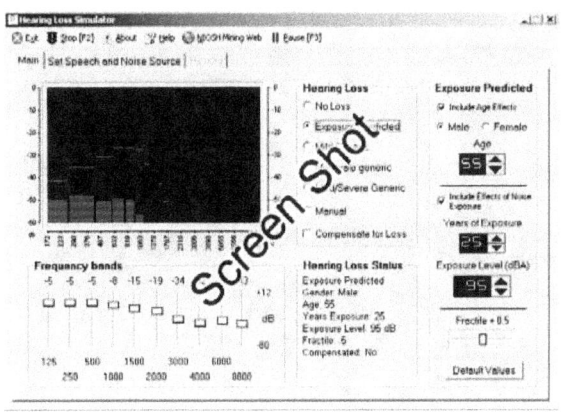

Equipment Needed

- Windows computer with speakers
- Hearing Loss Simulator software, which can be downloaded from: http://www.cdc.gov/niosh/mining/products/product47.htm

Instructions

Double-click on the Simulator. Once open, select "Exposure Predicted" under the "Hearing Loss" heading and click "Start" on the toolbar. Now adjust age, years of exposure, and decibel level settings to note how each affects the quality of sound emitted from the computer speakers. The Simulator allows the user to experience the hearing ability of people of different ages and exposure levels. Click on the "Set Speech and Noise Source" tab to change the type of noise.

For More Information

To request a Hearing Loss Simulator CD, contact:
Robert F. Randolph, NIOSH Pittsburgh Research Laboratory,
P.O. Box 18070, Pittsburgh, PA 15236
Phone: 412–386–4660; e-mail: RRandolph@cdc.gov

Topic: Hearing	Age Awareness Resources

Additional Resources

NIOSH Mining Safety and Health Topic – Hearing Loss:
http://www.cdc.gov/niosh/mining/topics/topicpage7.htm

NIOSH Mining Hearing Loss Prevention Workshops:
2005 – http://www.cdc.gov/niosh/mining/topics/hearingloss/hlprevworkshop/hlprevworkshop.htm
2006 – http://www.cdc.gov/niosh/mining/topics/hearingloss/hlprevworkshop2/hlprevworkshop.htm

NIOSH – Common Hearing Loss Prevention Terms:
http://www.cdc.gov/niosh/hpterms.html

Medline Plus – Hearing Disorders and Deafness:
http://www.nlm.nih.gov/medlineplus/hearingdisordersanddeafness.html

MSHA Health Talk – Noise Exposure for Surface and Underground Mines:
http://www.msha.gov/district/mnm/rmdist/rm%2Dhealthtalk.pdf

Washington University Hospitals, St. Louis, MO
Department of Otolaryngology: How Does the Ear Work?
http://wuphysicians.wustl.edu/dept.asp?pageID=14&ID=8

National Institute on Aging, Hearing Loss:
http://www.niapublications.org/engagepages/hearing.asp

National Institute on Deafness and Other Communication Disorders –
Hearing, Ear Infections, and Deafness:
http://www.nidcd.nih.gov/health/hearing

National Hearing Conservation Association:
http://www.hearingconservation.org/

Relevant Federal Regulations (intended as a guide, not a comprehensive list)

MSHA Health Standards for Occupational Noise Exposure:
http://www.msha.gov/1999noise/noise.htm

OSHA Safety and Health Topics – Noise and Hearing Conservation:
http://www.osha.gov/SLTC/noisehearingconservation/

Topic: Attention

Age Awareness News

Responding to People, Signals, and Events

Imagine you're operating a loader. You notice a dial that seems to have an unusual reading, and you also think you hear an alarm.

What goes through your mind? Do you respond to the control panel reading or the alarm you think you hear? In what order do you respond? How do you continue to operate the loader during your response to the other signals?

Our ability to respond effectively and correctly to these situations depends on the match between the control or alarm design and our own abilities. These abilities include perception, attention, and memory. When mismatches occur, safety or productivity can be affected.

Key Points:

- Older workers have vast amounts of corporate knowledge that should be used in everyday activities as well as training.

- Working memory loads can be reduced by providing lists or by "chunking" information.

- Tasks that require simultaneous monitoring should be presented in different ways (an auditory and a visual signal instead of two auditory signals).

How It Works

In order for us to respond to an event, we must first perceive it and then act upon it. For instance, we need to hear the alarm, know that it means to act a certain way, and then successfully perform that action. If it's an action that we've done many times, we can respond without directing much attention to it. The action is automatic.

Sometimes, however, we can't automatically respond to an event (listening to an announcement). We need to direct our attention toward it. Once we direct attention toward the signal or event, we use our thought and memory systems to process it and respond. It is important to realize that different types of events may have different demands on our attention. We may need to focus on only one thing (selective attention), we may be required to focus on two or more events (divided attention), or we may need to stay on constant alert (vigilance). If we are distracted or if there are many things happening, often we may not direct enough attention toward a particular event/object. We have all experienced this when too many people are trying to speak to us at once.

Memory is divided into two main types: short-term (working) or long-term. An example of short-term memory would be when someone asks you what you did earlier in the day.

Long-term memory is the storage of those items that are events in time, knowledge about words or concepts, or ways to do certain tasks, such as standard operating procedures (SOPs).

Normal Age Changes

Working memory and our ability to perform divided attention tasks decrease with age, but they can be helped through task redesign and assists. For instance, older people may have difficulty remembering new instructions that have many steps or focusing on several items at once. To assist workers in these tasks, provide them with a written checklist covering the steps necessary to complete the job.

However, when people have done a task repeatedly, the process becomes automatic. Automatic tasks are not as affected by aging as those tasks that require a lot of attention.

While we've mentioned systems that show declines with aging, we also want to emphasize that people gain expertise with age. All of us have benefited from a mentor or experienced that event at work requiring the senior person with years of knowledge to solve the problem!

Changing the Workplace to Reduce Risk

The workplace can be improved to lessen the effect of changes to a person's attention or memory.

For visual tasks, displays should be kept simple with clear distinctions between signal and background. Information should be presented in smaller "chunks," requiring less working memory. For instance, our phone numbers are broken down into three smaller "chunks" of numbers to make them easier to remember.

To reduce the load on working memory, provide materials for employees to write down instructions and expand SOPs.

Tasks can be redesigned to simplify the number of simultaneous signals or allow the signals to be presented in different formats (auditory and visual displays).

Controls should be designed to be intuitive (controls should move in the direction of the output). If multiple controls are housed together, different knob designs or handle lengths can be used so that the operator can easily distinguish between them.

Preventive Maintenance for People

- Regularly have your vision and hearing examined.

- Implement training programs designed to improve working memory and attention.

- Use previous task analyses and SOPs to reduce mental workload.

- Review your prescription drugs with a medical professional to determine if they may affect memory.

- Reduce personal stress, since stress can affect your memory and can cause you to be distracted.

- Good sleep habits can enhance your ability to process and recall information.

- Physical exercise has been shown to improve worker performance and also reduce stress. (When beginning a physical exercise program, you should always consult your physician.)

Common tasks that can overload our ability to respond to events or signals:

- *Monitoring two systems simultaneously*
- *Tasks that require large amounts of working memory, such as remembering procedures that are told to us rather than provided in written form*

Topic: Attention

| Topic: Attention | # Age Awareness Safety Talk Guide |

Introduction

- How we respond to people, signals, and events depends on our ability to perceive information, pay attention to it, remember it, and then act on it.
- Directing our attention to several things (divided attention) becomes more difficult as we age.
- Memory can either be short-term or long-term. A form of short-term memory called working memory decreases with age.
- Although divided attention and working memory can be negatively affected by age, expertise (also called corporate knowledge) increases with age.

Some Tasks Affected by Attention and Memory

- Divided attention: Tasks that require thinking about more than one thing at a time, for example, operating a continuous miner and monitoring the roof, or operating a front-end loader and looking out for pedestrians.

- Memory: Remembering the sequence of maintenance steps for nonroutine tasks. Getting directions for a task earlier in the day and remembering them later at the worksite.

Ways to Reduce Risk

- Make information, like that in task instructions, easier to remember by presenting it in smaller groups of five to seven items called chunks.
- Create and use a list of procedures for nonroutine tasks.
- Be aware that certain medications (for example, some sleeping pills, antidepressants, and antianxiety medications) are known to negatively affect memory. Check with your physician if you think your medication may be putting you at risk.
- Have your vision and hearing checked regularly. Problems with either will decrease your ability to respond to events.
- Improve your cognitive abilities and reduce stress with physical exercise. Be sure to consult your physician before beginning a new physical exercise program.

Older workers are valuable resources to companies because of their years of experience and expertise.

Divided attention and working memory are two processes that are known to be affected by aging.

Declines in either working memory or divided attention can be offset by modifications to the work environment.

What tasks at this worksite require a lot of attention and/or memory?

(For example: crusher operators, haulage truck drivers, continuous miner operators)

___

___

___

___

___

___

How can this worksite be changed to reduce attention- and memory-related risk?

(For example: removing visual "clutter," intuitive controls, less multitasking)

___

___

___

___

___

___

What can we do to improve our attention and memory?

(For example: make sure medications aren't affecting our work performance, regular exercise)

___

___

Topic: Attention

Topic: Attention

Age Awareness Activity

The Stroop Test

This test (designed by J. Ridley Stroop in the 1930s) shows how the brain processes information. It shows that while we can instantly read a word, it takes longer to name the color in which the word is written. The cognitive mechanism involved in this task is called inhibition; you have to inhibit, or stop one response and say or do something else.

Equipment Needed

- A stopwatch or a watch with time in seconds
- The Stroop chart provided on the next page. (NOTE: The chart must be shown in color.)
- Color printer or color computer monitor to display the chart

Setup Instructions

Print out or display the chart found on the next page in color.

Procedure

1. Time a trainee reading the words presented in the Stroop chart.
2. Record that time.
3. Time the same trainee saying the <u>color</u> in which each word is written.
4. Compare the two times. You should find that the time to read the written words is much shorter than the time needed to say the color in which each word is printed.

Why Is This Important?

The Stroop test emphasizes how people can have difficulty inhibiting previously known information or preferences. For instance, if a person is used to controls being configured in a certain way and then operates a piece of equipment where the controls are not organized that way, he or she may make mistakes. These types of mistakes can have serious consequences.

For More Information

Stroop effect: http://faculty.washington.edu/chudler/words.html

Stroop Chart

BLUE GREEN ORANGE

PINK RED BLACK PURPLE

TAN WHITE BROWN

Topic: Attention

Topic: Attention

Age Awareness Activity

Working Memory Exercise

This exercise demonstrates a technique that allows you to more easily remember lists.

Equipment Needed

- A stopwatch or a watch that shows seconds
- The list of sports words shown below

Skiing	Basketball	Tennis	Long Jump	Bobsledding
Diving	Hockey	Baseball	Ice Skating	Wrestling
Golf	High Jump	Volleyball	Lacrosse	Soccer
Swimming	Rugby	Cricket	Hurdles	Football

Procedure

1. Have trainees read over the list of sports for 30 seconds, then remove or cover the list.
2. Have the trainees write down as many items as they can remember.
3. Take a break for a few minutes.
4. Ask trainees to look at the original list of sports and try to mentally organize it into categories that make sense to them. (For example, the sports above could be put into categories of winter sports, summer sports, track and field events, etc.)
5. Give trainees another 30 seconds to read the list, reminding them to try to mentally categorize the items.
6. Remove or cover the list, and have trainees again write down as many items as they can remember.
7. Ask trainees to compare their two response lists. Ask them which is most accurate. It should be the second list.

Why Is This Important?

Although working memory can be affected by normal aging, techniques can be used to help improve memory for lists of items. The technique in this exercise demonstrates how "chunking" the information helped to better remember the words in the list.

For More Information

Age-related Changes in Memory by C. E. Barber:
http://www.ext.colostate.edu/pubs/consumer/10243.html

Memory Loss With Aging – What's Normal, What's Not:
http://familydoctor.org/124.xml

| Topic: Attention | Age Awareness Resources |

Additional Resources

Age-related Changes in Memory by C. E. Barber: http://www.ext.colostate.edu/pubs/consumer/10243.html

Charness N, Schaie KW, eds. [2003]. Impact of technology on successful aging. New York: Springer Publishing Co., Inc.

Fisk AD, Rogers WA, Charness N, Czaja SJ, Sharit J [2004]. Designing for older adults: principles and creative human factors approaches. Boca Raton, FL: CRC Press.

Memory Loss With Aging – What's Normal, What's Not: http://familydoctor.org/124.xml

National Research Council [2004]. Health and safety needs of older workers. Washington, DC: National Academies Press.

Topic: Musculoskeletal

Age Awareness News

The Pains and Strains of Sprains

Jobs in the mining industry can be physically demanding and provide many chances for sprain (ligament injury) and strain (muscle injury) to occur. Often, these injuries are the result of stress on tissues that increase over time.

However, as workers age, changes to the body may decrease the body's resistance to these types of injuries.

Muscular strength tends to decline with age. Ligaments and tendons weaken and become less flexible. These age-related changes may be why injuries to these tissues occur more often in older workers than in younger ones.

Much has been learned about how and why these injuries happen and the steps that can be taken to reduce injury risk to the musculoskeletal system (bones, muscles, tendons and ligaments). While eliminating such injuries entirely is impossible, there are many ways to reduce the risk of injuries to the aging miner.

Key Points:

- As people age:
  - Muscular strength declines.
  - Ligaments and tendons become less flexible.
  - Bones become weaker.

These changes lead to increased injury risk.

- Major risk factors that increase strain and sprain injury risk include:
  - Tasks requiring awkward work postures
  - Tasks requiring high force
  - Highly repetitive tasks

How It Works

The musculoskeletal system provides form and stability to the body and allows it to move.

Bones are constantly changing tissues that act as rigid structures of the body and as shields to protect delicate organs.

Joints are formed where bones come together. The configuration of a joint determines the degree and direction of possible motion. Some joints (like the knee) act as hinges; other joints (like the shoulder) are ball and socket and allow more complex movement.

Skeletal muscles, which are responsible for posture and movement, are attached to bones and arranged in opposing groups around joints. For example, muscles that bend the elbow (biceps) are countered by muscles that straighten it (triceps).

Tendons are tough bands of connective tissue mostly made of a protein called collagen. They firmly attach each end of a muscle to a bone and do not stretch.

Ligaments are strong, fibrous cords that connect one bone to another. They stretch to some extent. Ligaments surround, bind, strengthen, and stabilize joints.

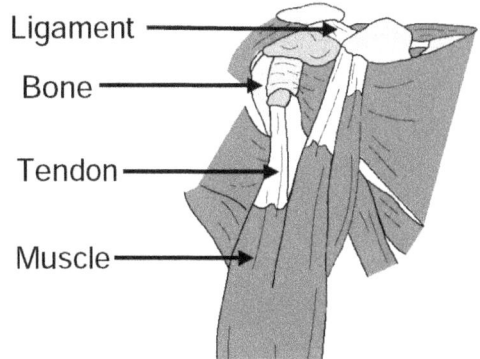

How Injuries Occur

Muscles can generate force both when they are shortening and lengthening. For example, in a weight lifter's "curl" exercise, the bicep muscle is shortening (and creating force) when the weight is raised. However, if the weight is slowly lowered, the biceps will create force while the muscle lengthens. Most muscle strains are the result of this type of lengthening contraction.

Overuse can cause tendons to become frayed or damaged, and the friction of tendons rubbing repeatedly against bone can lead to heat buildup. Both can lead to painful inflammation.

Ligaments can also become damaged if overstretched. This usually happens if the joints they stabilize are at extreme angles (awkward postures).

Topic: Musculoskeletal

Normal Age Changes

From about age 30, the density of bones begins to diminish in men and women. This loss of density accelerates in women after age 45. As a result, the bones become more fragile and are more likely to break, especially in women over 45.

As some people age, the surfaces of the joints do not easily slide over one other. This process can lead to osteoarthritis. Additionally, joints become stiffer because the connective tissue within ligaments and tendons becomes more rigid and brittle. This change also limits the range of motion of joints (i.e., reduces flexibility).

Aging is associated with a decrease in the amount of muscle in the body. The result is a gradual loss of muscle mass and muscle strength. Fortunately, the loss in muscle mass and strength can partially be overcome or at least significantly delayed by regular exercise.

Older injured workers also take longer to heal. These workers may need to avoid stressful tasks to heal properly.

Changing the Workplace to Reduce Risk

While the mining industry often requires high physical demands that can lead to strains and sprains, there are many steps that can be taken to reduce the risk.

Try to arrange tasks so that neutral postures can be used. Avoid bending, twisting and reaching. Storing materials at proper heights can help prevent bending and twisting.

Avoid repetitive or long-duration tasks by incorporating job rotation or rest breaks.

Proper work station (including seating and positioning of controls) design can improve posture and reduce injury risk.

Use power tools and hold parts with jigs or clamps to reduce the force required for a job.

Seats need to be adjustable so that they can support workers of different size properly.

Preventive Maintenance for People

The rate of decline in physical capabilities that occur with aging can be slowed with proper care and exercise. Tissues of the body respond to the demands placed on them. If no stress is present, the tissues weaken. A moderate amount of stress will make tissues stronger. The following are practices that can help maintain a healthy musculoskeletal system:

- Regular weight-bearing exercise is a good way to maintain bone density and is particularly important for women after age 45. Calcium supplements are helpful for bone health and may be most useful for older women.

- Aerobic exercise helps the musculoskeletal system by strengthening muscles, bone, and cartilage, which provides protection against future injuries.

- A stretching program can help maintain flexibility. This may be one of the most underrated measures for preventing injury.

| Topic: Musculoskeletal | Age Awareness Safety Talk Guide |

Stan is a 52 year-old worker in the rail division of a large silica sand mine. One of his jobs is to open gates underneath the covered hopper railcars to clean the cars out. He would take a long steel breaker bar into the gate opening crank and forcefully rotate it repeatedly. Stan was always feeling pain in his shoulders, elbows, and wrists. Last winter, he injured his shoulder when trying to force open a frozen gate.

A friend told him about a hydraulic powered gate opener. The device could be wheeled into position, attached to the crank, and used to open the gates for cleaning. Stan's supervisor agreed to purchase the gate opener. Since making the switch, Stan's pains have reduced.

Introduction

- The musculoskeletal system is the body's system of bones, muscles, tendons, and ligaments.
- Jobs that require excess strength, repeated motions, or awkward postures can result in sprain and strain injuries.
- These injuries are usually the result of stress on the muscles and other tissues that develop over time.
- The bones become more fragile and are more likely to break, especially in old age.
- As workers age, changes to the body may decrease the body's resistance to these types of injuries.

Some Tasks Affected by Changes to the Musculoskeletal System

- Lifting, pushing, pulling, or carrying objects (especially heavy objects)
- Use of handtools
- Any work that involves exerting force in an awkward posture

Ways to Reduce Risk

- Arrange work so body joints are in a neutral posture. Avoid bending and twisting. Organize the work area to avoid lifting or working above shoulder level or below knee level.
- Choose the correct tool for the job. When using handtools, choose a tool that allows the wrist to remain in a neutral posture. Sometimes a bent tool handle can help maintain a straight wrist.
- Reduce force demands of the task. Power tools usually require less force than non-power tools. Using jigs and clamps to hold parts being worked on requires less force than holding parts in your hand.
- Avoid repetitive tasks by incorporating job rotation or rest breaks.
- Design for adjustability. When possible, use components (seats, controls, viewing angles for monitors) that can be adjusted to individual operators.

What are some tasks with musculoskeletal risks at this worksite?
(For example: jobs requiring strength, repeated motions, awkward postures)

_____
_____
_____
_____
_____

How can this worksite be changed to reduce musculoskeletal risk?
(For example: adjusting posture, use of power tools, avoiding repetitive tasks)

_____
_____
_____
_____
_____

What can we do to reduce risk of injury to our musculoskeletal system?
(For example: weight-bearing exercise, maintain healthy body weight, aerobics)

_____
_____

Topic: Musculoskeletal

| Topic: Musculoskeletal | **Age Awareness Activity** |

Reducing Carpal Tunnel Syndrome (CTS)

The carpal tunnel is located in the wrist and is formed by several bones and a ligament (see Figure 1). Nine tendons that flex your fingers, such as when grasping a ball, pass through the carpal tunnel, along with the median nerve. When the space within the carpal tunnel is reduced, the median nerve is squeezed and may result in CTS.

Equipment Needed

- Nine 3/16- by 5-inch wooden dowels
- Nine 1/4- by 5-inch wooden dowels
- One 1-inch round PVC pipe
- One 1/2-inch round shrink tubing

Procedure

1. Place the shrink tubing (median nerve) and nine small (3/16-inch) dowels (tendons) inside the pipe (Figure 2).
2. Explain that when the tendons are healthy, there is enough space such that the tendons can move through the tunnel easily and the median nerve is not compressed.
3. Replace the small dowels with the larger (1/4-inch) dowels (Figure 3).
4. Explain that these large dowels represent the swollen tendons.
5. Point out that the median nerve is compressed because the swollen tendons take up more space in the compact carpal tunnel.
6. Explain that this compression of the nerve is responsible for the symptoms of CTS.

Why Is This Important?

CTS has been associated with certain risk factors, such as forceful exertions, repetition, and awkward postures of the wrist (wrist flexion, extension, and deviation). The risk for CTS increases when more than one of these risk factors is present. These risk factors may directly reduce the size of the carpal tunnel or result in inflammation or irritation of the tendons, which increases the size of the tendons. In either situation, the space available within the carpal tunnel for the median nerve is reduced. If the nerve is squeezed over a prolonged period of time, changes to the nerve occur and pain and discomfort may result. If these risk factors are present in tasks that you do, consider ways to change the task to reduce or eliminate the risk factor exposures.

For More Information

Putz-Anderson V, ed. [1988]. Cumulative trauma disorders: a manual for musculoskeletal disease of the upper limbs. London: Taylor and Francis.

OSHA Ergonomic Principles Index: http://www.osha.gov/SLTC/etools/electricalcontractors/supplemental/principles.html

Activity Graphics

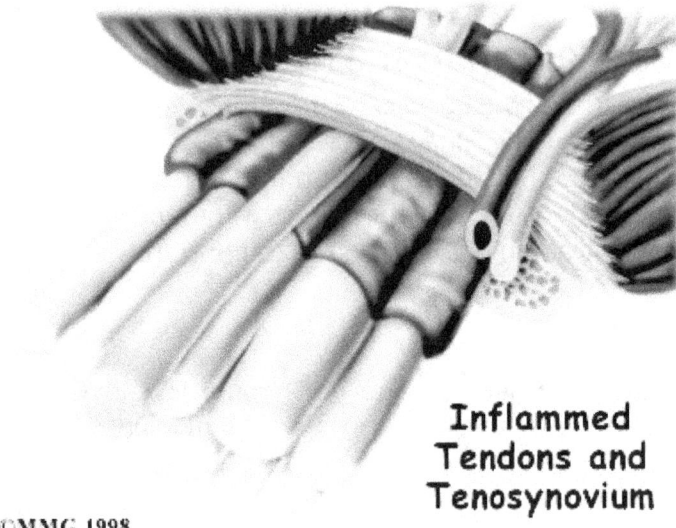

Figure 1.—Graphics showing the location of the carpal tunnel and what happens when inflammation occurs.

Inflammed Tendons and Tenosynovium

Transverse carpal ligament

Carpal bones

Carpal Tunnel Opening

Figure 2.—Nine small (3/16-inch) dowels and shrink tubing.

Figure 3.—Nine large (1/4-inch) dowels and shrink tubing.

Topic: Musculoskeletal

| Topic: Musculoskeletal | Age Awareness Resources |

Additional Resources

Dul J, Weerdmeester BA [2001]. Ergonomics for beginners: a quick reference guide. 2nd ed. London: Taylor and Francis.

Putz-Anderson V, ed. [1988]. Cumulative trauma disorders: a manual for musculoskeletal disease of the upper limbs. London: Taylor and Francis.

NIOSH Mining – Musculoskeletal Diseases: http://www.cdc.gov/niosh/mining/topics/topicpage8.htm

NIOSH – Ergonomics and Musculoskeletal Disorders: http://www.cdc.gov/niosh/topics/ergonomics/

OSHA Safety and Health Topics – Ergonomics: http://www.osha.gov/SLTC/ergonomics/

Relevant Federal Regulations (intended as a guide, not a comprehensive list)

MSHA does not have specific regulations that it uses to cite for undue physical demands.

Operations that fall under OSHA jurisdiction may be cited under the "General Duty Clause" (Section 5 of the Occupational Safety and Health Act):

(a) Each employer --
  (1) shall furnish to each of his employees employment and a place of employment which are free from recognized hazards that are causing or likely to cause death or serious physical harm to his employees;
  (2) shall comply with occupational safety and health standards promulgated under this Act.

(b) Each employee shall comply with occupational safety and health standards and all rules, regulations, and orders issued pursuant to this Act which are applicable to his own actions and conduct.

OSHA cites for ergonomic disorders under the provisions of paragraph 5(a)(1). Although the language in paragraph 5(b) holds employees responsible for complying with health and safety standards, the employer bears most of the responsibility for compliance in the eyes of OSHA.

OSHA inspectors may issue a citation under the General Duty Clause when the following criteria are met:

- There is not an applicable OSHA standard.
- The employer failed to keep the workplace free of a hazard to which employees of that employer were exposed.
- The hazard is (or should have been) recognized by the employer.
- The hazard is causing or was likely to cause death or other serious physical harm.
- There is a feasible and useful method to correct the hazard.

The absence of specific ergonomic standards requires interpretations when using the General Duty Clause [Ergoweb 2008].

Topic:
Lower Back

Age Awareness News

Your Aching Lower Back

Key Points:

- The process of aging (in combination with physical loading) causes degeneration of spine tissues.

- Back pain is often the result of "wear and tear" that occurs due to the combination of aging and high loads experienced by the spine.

- The spine can experience *half a ton of force* during heavy lifting tasks.

- Bending over when lifting can triple the forces on the spine.

- Lifting jobs can often be easily redesigned to reduce the load on the spine.

Low back pain will affect about four out of every five people at some point during their lives. This pain can greatly influence activities of daily living and is a major cause of disability and lost wages among workers.

A typical time course for those who experience back pain is to first experience it in your early to mid-twenties, sometimes as the result of a heavy exertion such as lifting. Once an episode has occurred, it is likely that additional episodes will be experienced.

Episodes of pain generally increase in frequency in the thirties and forties. Herniated discs occur most often between the ages of 45 and 55. Symptoms often lessen after age 60.

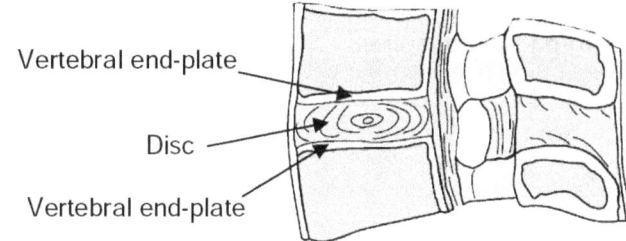

How It Works

As seen in the picture on the next page, the spine is made up of a series of 24 bones called vertebrae, which are divided into four major regions. The lower portion of the spine (lumbar spine) is the part of the spine that usually experiences the highest stresses and is a common site for pain.

Discs are ligamentous structures that separate the vertebrae and increase the flexibility of the spine.

There are many possible causes of back pain. Recent research has suggested that disc degeneration in the spine may be a frequent cause, especially of chronic (persistent) back pain.

Discs do not have a blood supply. Their health is dependent on nutrition from the vertebrae. Nutrients from blood vessels in the bone travel through the vertebral end-plates into the discs.

Unfortunately, the vertebral end-plates seem to be the weakest structure in the spine and can fracture when heavy loads are placed on the spine (as in lifting tasks).

While these end-plate fractures can be repaired, they heal by forming scar tissue, which reduces the flow of nutrients to the disc. As nutrition is reduced, the disc starts to degrade. Cracks (or fissures) will then begin to develop from the inner to the outer portion of the disc.

These cracks may allow some of the gel in the center of the disc to leak outward. If this gel reaches the outer portions of the disc, it may contact pain fibers, resulting in low back pain.

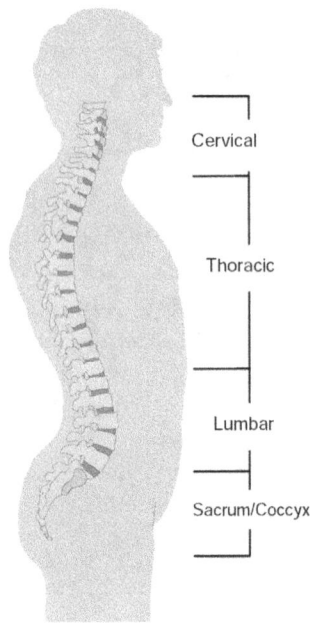

Regions of the spine.

Common tasks that lead to low back pain:
- *Lifting heavy objects, especially if bending to lift from the floor*
- *Pushing and pulling*
- *Driving if exposed to vibration, jolts, or jarring*

Topic: Lower Back

Normal Age Changes

The natural effects of aging on the low back include osteoporosis (a decrease in the amount of bone), decreased strength and elasticity of muscles, and decreased elasticity and strength of ligaments.

Bone loss in the vertebrae of the spine can be a particularly serious problem for women over the age of 45. However, men can also have a great deal of bone loss with age. The result of bone loss is weaker vertebrae that are more likely to fracture.

As described earlier, endplate fractures start the process of disc degeneration. With age, almost everyone experiences some disc degeneration. However, some people will have minor degeneration, while others will have severe changes.

Aging also affects muscles, tendons, and ligaments that support the spine. A decline in muscle strength is experienced after age 30. With this change, muscles are less able to withstand the high loads placed on the spine during activities such as lifting, pushing, and pulling. Tendons and ligaments generally become weakened and less flexible with age, which increases the chance that these tissues may become injured.

Changing the Workplace to Reduce Risk

Workplaces can often be redesigned to reduce the risk of back pain. The following are some simple methods to improve the design of tasks to reduce back pain risk:

Use mechanical-assist devices (such as hoists or jacks) to reduce demands of heavy and awkward lifts where feasible.

Organize work so that lifting and carrying of items are minimized.

Relocating items from floor to waist level (storing on tables or benches) can greatly reduce stress on the low back when lifting.

Remove barriers that prevent access to needed materials and tools.

Always keep the load close to the body when lifting or carrying.

Organize work so that periods of lifting are interspersed with periods of rest.

Organize work so that physically demanding work is done later in the day.

Avoid bending the trunk, especially at the beginning of a shift.

Awkward postures (stooping, kneeling, squatting, sitting) result in lower lifting capacity. Try to keep loads light when using these postures.

Preventive Maintenance for People

Although the normal effects of aging that result in decreased bone mass, strength, and elasticity of muscles and ligaments can't be avoided, they can be slowed by:

- Exercising regularly to keep the back and abdominal muscles that support the spine strong and flexible
- Using the correct lifting and moving techniques; get help if an object is too heavy or an awkward size
- Maintaining proper body weight; being overweight strains back muscles
- Avoiding smoking, which may also increase risk of low back pain
- Maintaining proper posture when standing and sitting—don't slouch

Topic: Lower Back

Age Awareness Safety Talk Guide

Before each batch of bagging at a minerals processing plant, Charlie (age 48) lifted several 50- or 100-pound bags from a conveyor to a scale on the floor.

One day he bent forward to lift a 100-pound bag and suddenly felt severe pain in his lower back. He was off work for 2 weeks.

After his injury, the job was reviewed and the scale was moved to a platform level with the conveyor. The change will reduce the risk of back injury.

Why wasn't the job changed before Charlie got hurt?

Introduction

- Low back pain will affect about four out of every five people at some point during their lives.
- Back pain can start early in life (typically in your twenties) and increases in your thirties and forties. Symptoms sometimes lessen after age 60.
- The effects of normal aging on the low back are:
  - Osteoporosis (decreased amount of bone)
  - Decrease in strength and elasticity of muscles, ligaments, and tendons
- Physically demanding jobs increase the risk of low back pain.
- Designing jobs to reduce the stress on the low back can reduce the risk of low back pain.

Some Tasks Affected by Low Back Pain

- Lifting heavy objects, especially if bending to lift from the floor
- Pushing and pulling
- Driving if exposed to vibration, jolts, or jarring

Ways to Reduce Risk

Job changes can often be made easily and inexpensively.

- Use mechanical assists for heavy loads whenever possible.
- Design tasks to avoid bending forward to lift. If possible, store items to be lifted between knee and shoulder height, *not on the floor*.
- Keep the load close when lifting. The farther away the load is from the body, the greater the stress on the back. Remove any barriers that prevent workers from getting close to the load.
- Get help if an object is too heavy or is an awkward size.
- Keep thinking of ways that you can organize your work to reduce lifting demands.

What are some tasks that may have low back pain risks at this worksite?

(For example: overhead lifts required, heavy lifts, awkward postures while performing maintenance tasks)

_____

_____

_____

_____

_____

How can this worksite be changed to reduce low back pain risk?

(For example: use mechanical assists, remove unnecessary barriers, use two people for awkward lifts)

_____

_____

_____

_____

_____

What can we do to reduce our risk of low back pain?

(For example: increase weight-bearing exercise, flexibility)

_____

_____

Topic: Lower Back

| Topic: Lower Back | # Age Awareness Activity |

Keeping the Load Close

The distance between the load and the body is an important factor in the stress on the body when lifting. This exercise shows how the difficulty of a lift changes with how far the object is from the body.

Equipment Needed

- A light weight (3–5 pounds)
- A stopwatch

Procedure

Give a trainee the weight and the following instructions:

1. Place the weight in your hand with your arm straight out in front of you and your elbow fully extended. Keep your elbow slightly bent so that you do not lock out your joint.
2. Hold this posture for 30 seconds to 1 minute.
3. Now bend the elbow so that the hand holding the weight is close to the body.
4. Again hold this posture for 30 seconds to 1 minute.

Ask which posture made it easier to hold the weight. It should be much easier to support the weight when it is closer to the body.

Why Is This Important?

Posture has a major effect on strength and the stress on the body when lifting. As you can see and feel from the exercise, the weight seems "lighter" when it is held closer to the body. Workers of every age should be reminded of the importance of keeping the load close to the body when lifting.

For More Information

Dul J, Weerdmeester BA [2001]. Ergonomics for beginners: a quick reference guide. 2nd ed. London: Taylor and Francis.

OSHA Ergonomic Principles Index: http://www.osha.gov/SLTC/etools/electricalcontractors/supplemental/principles.html

Topic: Lower Back

Age Awareness Activity

Wear and Tear on the Back

Many back injuries are thought to be the result of damage that builds up over time. As the back experiences loading (for example, when lifting objects), small cracks can appear in the end-plates. With repeated loading, these small cracks get larger and eventually can be severe enough that nutrition to the disc is impeded. This will lead to disc degeneration, which is a common cause of back pain.

Equipment Needed

A paper clip or a metal coat hanger.

Procedure

If using paper clips, try to bring enough for everybody. If using a coat hanger, you might need to demonstrate by yourself. Have the trainees straighten out the paper clip, then ask them to bend the paper clip (or bend the hanger) back and forth and have them count the number of times they bend it before it breaks in two.

Why Is This Important?

Like many materials, the spine can get damaged by a process of repeated loading. Just like the paper clip (or hanger), repeated loading eventually causes an area of weakness to develop and a small crack will form. This area of weakness, when subjected to additional loading cycles, will expand further and further until a complete failure of the material occurs. People tend to believe that if they can lift a certain weight one time without injury, they are safe in lifting it many times. However, in reality, small amounts of damage may be accumulating that can result in an injury. For this reason, lifting tasks should be designed to minimize spine loading. This can often be done by use of mechanical-assist devices or by better design of lifting tasks (to reduce the load experienced by the spine).

For More Information

Gallagher S, Marras WS, Litsky AS, Burr D [2005]. Torso flexion loads and the fatigue failure of human lumbosacral motion segments. Spine *30*(20):2265–2273.

| Topic: Lower Back | Age Awareness Resources |

Additional Resources

Dul J, Weerdmeester BA [2001]. Ergonomics for beginners: a quick reference guide. 2nd ed. London: Taylor and Francis.

Kroemer KHE [1997]. Ergonomic design of material handling systems. Boca Raton, FL: CRC Press, LLC.

Chaffin DB, Andersson GBJ [1991]. Occupational biomechanics. 2nd ed. New York: John Wiley and Sons, Inc.

NIOSH Mining – Musculoskeletal Diseases: http://www.cdc.gov/niosh/mining/topics/topicpage8.htm

NIOSH – Ergonomics and Musculoskeletal Disorders: http://www.cdc.gov/niosh/topics/ergonomics/

OSHA Safety and Health Topics – Ergonomics: http://www.osha.gov/SLTC/ergonomics/

Relevant Federal Regulations (intended as a guide, not a comprehensive list)

MSHA does not have specific regulations that it uses to cite for undue physical demands.

Operations that fall under OSHA jurisdiction may be cited under the "General Duty Clause" (Section 5 of the Occupational Safety and Health Act):

(a) Each employer --
- shall furnish to each of his employees employment and a place of employment which are free from recognized hazards that are causing or likely to cause death or serious physical harm to his employees;
- shall comply with occupational safety and health standards promulgated under this Act.

(b) Each employee shall comply with occupational safety and health standards and all rules, regulations, and orders issued pursuant to this Act which are applicable to his own actions and conduct.

OSHA cites for ergonomic disorders under the provisions of paragraph 5(a)(1). Although the language in paragraph 5(b) holds employees responsible for complying with health and safety standards, the employer bears most of the responsibility for compliance in the eyes of OSHA.

OSHA inspectors may issue a citation under the General Duty Clause when the following criteria are met:

- There is not an applicable OSHA standard.
- The employer failed to keep the workplace free of a hazard to which employees of that employer were exposed.
- The hazard is (or should have been) recognized by the employer.
- The hazard is causing or was likely to cause death or other serious physical harm.
- There is a feasible and useful method to correct the hazard.

The absence of specific ergonomic standards requires interpretations when using the General Duty Clause [Ergoweb 2008].

Topic: Work Capacity

Age Awareness News

Work Capacity and Endurance

Physical changes that occur as people age can affect the amount of effort workers can put forth (work capacity) and can decrease the amount of time they can perform physically demanding tasks (endurance).

Research indicates that the effects of aging on physical work capacity and endurance should be considered in the workplace.

Key Points:

- As people age, their ability to perform physically demanding work declines.

- Fatigue will occur more quickly for older workers and can lead to higher risk for injuries and accidents.

- Designing proper work-rest schedules can help prevent fatigue in older workers.

- Aerobic exercise can help prevent or minimize the decrease in work capacity that occurs with age.

Work capacity is highest in the late teens and early twenties and then decreases with age.

Work capacity declines more rapidly for individuals who do not adhere to a healthy diet and who do not stay physically active.

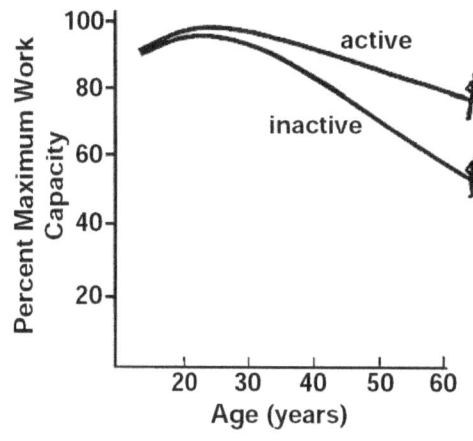

How It Works

Work capacity and endurance are affected by several systems in the body, including the heart and blood circulation, breathing, and the muscular system. Muscles need oxygen to perform work, and this oxygen is obtained from the air by breathing it into the lungs. Once in the lungs, the oxygen is transferred to the blood where it attaches to a molecule called hemoglobin. As the heart beats, the oxygen (attached to the hemoglobin) gets pumped through the arteries of the body and then into smaller blood vessels called capillaries. These capillaries are present everywhere throughout the body and help to deliver the necessary oxygen to body tissues. All cells in the body require oxygen. However, muscle cells put a very high demand on the oxygen delivery system, especially when performing heavy work. This is why, when you climb a flight of stairs, you breathe more heavily. The work performed by the leg muscles when climbing stairs (or in other physical tasks) demands an increased supply of oxygen. This signals the body to breathe in more air (breathing deeper and heavier) and pump blood more quickly (heart beats faster) so that the muscles can continue working. If you stop at the top of the stairs, the demand for oxygen by the muscles decreases, and your heart and breathing rates will gradually return to normal.

Terry's Tiring Task

Terry is a 55-year-old miner who has been doing belt cleanup for years. There was a time when he was known for making quick work of any belt cleanup task.

Recently, Terry has noticed he gets much more out of breath and has to take frequent breaks when shoveling the belt.

In his twenties, Terry was physically fit and regularly played sports. Lately, his only sports are the ones he watches his son play.

Over time, Terry's body has been changing. His work capacity is not what it used to be.

Terry can still do his job, but he may need a little more time to complete physically demanding tasks. He will need to take more rest breaks to keep from overexerting himself. He should get someone to help out on difficult tasks.

Topic: Work Capacity and Endurance

Normal Age Changes

After men and women reach full physical maturity (usually in their early to mid-twenties), a steady decline is seen in the ability to breathe in and transport oxygen to working muscles.

The degree of the decline seems to depend on the amount of aerobic exercise (or lack of it) that individuals get. Exercise can have a dramatic effect on the capacity of the oxygen transport system of the body.

Without sufficient aerobic exercise, the oxygen transport system can lose 25% of its capacity *in just 4 years!*

Changes in the muscular system can also be dramatic after age 45. In a 10-year period, strength capabilities can decrease up to 50%, and the ability of the muscle to get oxygen from the blood is also reduced.

Interestingly, these declines seem to be similar in people who work in both blue-collar and white-collar occupations. This indicates that "exercise" obtained from physically demanding jobs does not protect workers against declines in muscular strength and capacity. Regular physical exercise to maintain an adequate level of fitness is still necessary for workers in such jobs.

Changing the Workplace to Reduce Risk

International recommendations suggest that workers shouldn't exceed a workload of more than 50% of the maximum oxygen transport capacity of their bodies. Since there is a significant decline in work capacity with aging, it is important to reduce the physical requirements of jobs as much as possible due to the aging workforce.

Ways to reduce workload include flexible working times and allowing appropriate work-rest schedules to permit recovery to occur.

Training methods tailored to the needs of older workers can also reduce the physical demands on the aging workforce.

Preventive Maintenance for People

Regular aerobic exercise is critical to preventing loss of work capacity. In conjunction with a healthy diet, it can also help to avoid the numerous health risks associated with obesity.

Aerobic exercise includes any activity that uses large muscle groups and that can be maintained continuously for at least a 20- to 30-minute period. The following exercises can help maintain a good work capacity:

- Biking, running, cross-country skiing, swimming, walking briskly, and dancing
- Exercise machines like treadmills, elliptical trainers, and rowing machines can provide a good aerobic workout.
- Resistance training can increase muscular endurance, is low-impact, and provides many health benefits.

Finally, it is important to avoid smoking to preserve good lung function and maintain a healthy work capacity.

| Topic: Work Capacity | **Age Awareness Safety Talk Guide** |

Introduction

- Aging causes a gradual reduction in the amount of physical effort that workers can sustain without fatigue.
- Declines in work capacity and endurance generally start in early adulthood (mid-twenties).
- Decreased work capacity and endurance are due to reduced ability to:
    - Pump blood as quickly;
    - Breathe as deeply; and
    - Use oxygen efficiently (for example, in muscle).
- The reduced endurance to physically demanding tasks results in more rapid fatigue.
- Fatigued workers are more likely to make mistakes or become injured.

Some Tasks Where Work Capacity Is a Concern

- Prolonged or repeated lifting or pushing/pulling activities
- Shoveling (belt cleanup, etc.)
- Climbing stairs or ramps
- Carrying heavy materials long distances

Ways to Reduce Risk

- Design the workplace to reduce energy demands.
    - (For example, redesign lifting/carrying tasks to lighten the load or shorten the distance carried.)
- Provide mechanical aids or carts to move materials.
- Schedule less physically demanding tasks in between tasks that require heavy exertion.
- Schedule appropriate rest breaks during demanding tasks.
- At home:
    - Participate in aerobic exercise.
    - Avoid smoking.
    - Maintain a healthy body weight.

Which tasks require high work capacity or endurance at this site?
(For example: shoveling, carrying heavy materials long distances)

_____

_____

_____

_____

_____

How can this site be changed to reduce the risk of fatigue or injury?
(For example: use vehicles to move materials, provide rest periods or periods of lighter work)

_____

_____

_____

_____

_____

What can we do to reduce the age-related decline in work capacity?

(For example: do more aerobic exercise, stop smoking)

_____

_____

Topic: Work Capacity and Endurance

Topic: Work Capacity

Age Awareness Activity

Effect of Age on Work Capacity

The purpose of this activity is to demonstrate how aging affects the percentage of maximum work capacity at which a worker is operating. The decrease in maximum heart rate as workers age causes older workers to be at a higher percentage of their maximum work capacity for a given task.

Equipment Needed

- A watch with a second hand or a stopwatch
- Optional: a calculator and a heart rate monitor

Procedure

1. Have a volunteer perform a physical task such as shoveling, jogging in place, or climbing stairs for 60 seconds.
2. If available, have the volunteer wear a heart rate monitor.
3. Immediately after stopping, record the heart rate on the heart rate monitor or have the trainee take his or her pulse rate.

    For pulse rate, have the volunteer gently press two fingers on the neck underneath the angle of the jawbone. Count the number of pulse beats in 10 seconds (timed by a watch or stopwatch). Multiply this number by 6 for the total number of beats per minute.

4. Compare the heart rate to the maximum heart rate (220 minus the worker's age) for a 25-year-old (195 beats per minute) and a 65-year-old (155 beats per minute).
5. Do this by dividing the pulse rate by 195 for the maximum for a 25-year-old and by 155 for a 65-year-old.
6. For example, for a pulse rate of 140, the percentage of maximum heart rate for younger and older workers would be as follows:
    a. For a 25-year-old: 140/195 × 100 = 72%.
    b. For a 65-year-old: 140/155 × 100 = 90%.
7. Since the older worker would be at a higher percentage of maximum work capacity, this worker will fatigue more quickly and will not be able to continue the work as long.

Why Is This Important?

As workers age, they will generally have lower maximal heart rates and may fatigue more quickly. This means they may not be able to perform physically demanding jobs for as long as they did when they were young or may need to take additional time to complete a demanding task. However, other factors may influence this relationship. For example, older non-smokers may have a higher work capacity than younger smokers.

For More Information

McArdle WD, Katch FI, Katch VL [1981]. Exercise physiology: energy, nutrition and human performance: Philadelphia, PA: Lea & Febiger.

Activity Worksheet

Step 1: Get volunteer's heart rate

Number of heart beats in 10 seconds: [ ]

Multiply by 6:   × 6

Result (beats per minute): = [ ]

Step 2: Calculate % maximum work capacity (MWC) for a 25-year-old

Divide result from step 1 by 195 to get % MWC for a 25-year-old:

[ ] /195 = [ ] % MWC

Step 3: Calculate % maximum work capacity (MWC) for a 65-year-old

Divide result from step 1 by 155 to get % MWC for a 65-year-old:

[ ] /155 = [ ] % MWC

Note that the same amount of physical exertion results in a higher % MWC for the 65-year-old. The older worker will thus fatigue more quickly than the younger worker and may need additional recovery time.

Beats counted in 10 seconds	Beats per minute	MWC for 25-year-old	MWC for 65-year-old	Beats counted in 10 seconds	Beats per minute	MWC for 25-year-old	MWC for 65-year-old
15	90	46%	58%	23	138	71%	89%
16	96	49%	62%	24	144	74%	93%
17	102	52%	66%	25	150	77%	97%
18	108	55%	70%	26	156	80%	101%
19	114	58%	74%	27	162	83%	105%
20	120	62%	77%	28	168	86%	108%
21	126	65%	81%	29	174	89%	112%
22	132	68%	85%	30	180	92%	116%

Topic: Work Capacity

Topic:
Work Capacity

Age Awareness Resources

Additional Resources

National Research Council [2004]. Health and safety needs of older workers. Washington, DC: National Academies Press.

Astrand PO, Rodahl K [1977]. Textbook of work physiology: physiological bases of exercise. New York: McGraw-Hill.

Fleg JL, Morrell CH, Bos AG, Brant LJ, Talbot LA, Wright JG, Lakatta EG [2005]. Accelerated longitudinal decline of aerobic capacity in healthy older adults. Circulation *112*(5):674–682.

Topic: Falls

Age Awareness News

Slips, Trips, and Falls (STF)

There is a high risk of slipping and falling at mine sites because the ground or other walking surface is often uneven, wet, and/or frozen.

Approximately 25% of all musculoskeletal injuries sustained by mine workers are attributed to STF.

As shown in the graph below, older workers have higher percentages of injuries due to slips and falls than younger workers. Older miners also generally take longer to recover from injuries. This may be because of factors attributed to the aging process, such as loss of flexibility and lower-extremity strength.

Injuries from falls may be severe, and once they occur there can be a future predisposition to similar injuries.

Slips and falls are largely preventable through a proactive approach to workplace safety. Protect yourself and your coworkers from these potentially harmful and expensive injuries.

Key Points:
- For all industries, falls account for 12.5% of fatal occupational injuries.
- Falls on the same level account for 19.9% of all nonfatal occupational injuries involving days away from work.

(Bureau of Labor Statistics, 2003)

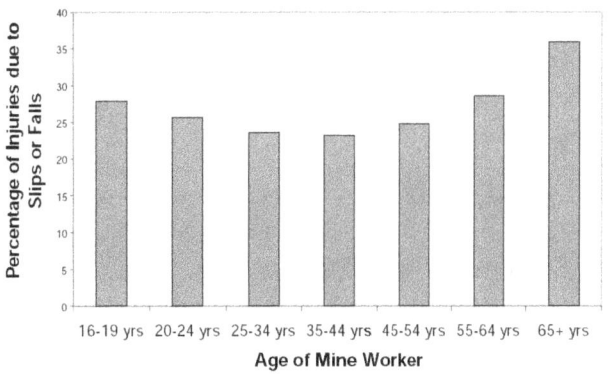

Percentage of Slip or Falls Injuries in Age Groups

How Does It Happen?

Slips and falls can be caused by a variety of individual and environmental factors, such as loss of coordination, loss of balance, fatigue, slippery surfaces, transitions in floor surfaces, uneven surfaces, cold weather, and poor housekeeping.

Although we can't always point to the cause of slips, there is usually a loss of traction between the shoe and the floor surface that leads to a foot slide and then possibly to a fall. Slipperiness of walking surfaces has been estimated to account for 40%–50% of all fall-related injuries.

Falls can occur while walking, getting on or off of equipment, or moving from a higher level to a lower level on steps or a slope.

Tips to reduce the risk of injuries:

- Check and maintain all walkways and access ladders.

- Wear appropriate, well-fitting footwear.

- Consider risk for falls when starting new jobs or in new environments.

- Take your time when getting on and off of equipment.

- Post warning signs around areas that may be slippery.

Normal Age Changes

The higher incidence and severity of falls in older workers can be related to several physiological changes.

As people age, their reaction times become slower and they may be unable to recover from a slip. When a slip occurs, a person needs to react quickly to recover balance and prevent a fall.

As people age, they also tend to experience declines in their lower-extremity strength before they experience those same changes in upper-extremity strength. With reduced muscle strength, a person may not be able to recover from a slip.

In addition, older workers may be less flexible and therefore unable to reach steps or handholds.

When older workers are injured, it is commonly a result of a fall to the floor, walkway, or ground.

Slips and falls as a percentage of total injuries is related to the age of the workforce. Miners at age 65 and older experience about a 50% increase in the number of slips and falls compared with 25- to 34-year-olds.

Additionally, older workers take longer to recover from falls than younger workers. Days lost for miners with STF injuries nearly quadruple from the 25–34 age group to the 55–64 age group.

Changing the Workplace to Reduce Risk

Make sure that walkways are properly deiced and free from excess water or other slippery materials.

Sweep up loose debris and inspect pathways for unexpected holes or uneven payment.

Be sure that work boots are not worn thin on the treads.

Be sure that steps on stairways and steps leading onto equipment are not covered with slippery materials such as caked mud or ice.

Use three-point contact while climbing up and down ladders.

Be sure that handrails are not broken and can be used.

Provide enough lighting so that uneven or unexpected ground changes can be seen.

Do not carry objects that obstruct your view of the ground.

Be sure that materials used to clean floors do not make them slippery.

Preventive Maintenance for People

- Regular exercise may help to retain balance throughout your lifetime.

- If you sense difficulties with balance, contact your physician to determine if it's related to a medical condition.

- Your vision should be checked regularly and corrective glasses worn if needed.

- Be sure that you are aware of any effects from medication that could reduce your alertness or balance.

Topic: Falls

| Topic: Falls | Age Awareness Safety Talk Guide |

Slips, trips, and falls are common, dangerous, and costly. The annual direct cost of fall-related occupational injuries in the United States has been estimated at $6 billion.

Slips, trips, and falls are preventable to a large degree because risk factors for these are known and can be addressed.

Introduction

- Slips can be caused by loss of coordination, loss of balance, fatigue, slippery surfaces, changes in floor surface, uneven surfaces, cold weather, and poor housekeeping.
- When a person slips, he or she must be able to respond correctly to reestablish balance in order to prevent a fall.
- Since the terrain at mine sites changes and is usually wet, frozen, or uneven, the risk of slipping and then falling is high.
- In general, older miners are more likely to have slip, trip, and fall-related injuries than younger ones.

Some Activities With Higher Risks for Falls

- Walking on icy ground
- Walking in poorly lit areas or over uneven ground
- Getting on and off of equipment
- Climbing or descending stairs and ladders

Ways to Reduce Risk

Assess hazards in your work area.
- Look for unexpected holes and uneven payment. Make repairs before someone falls.
- Be sure it's bright enough to see uneven ground.

Keep up with general housekeeping tasks.
- Ensure floors and walkways are free from water or other slippery materials. Properly deice areas when needed.
- Designate specific pathways and keep them clean from work debris.

Take extra care on equipment.
- Remove mud or ice from stairways or steps.
- Use three-point contact while getting on and off of equipment.
- Ensure handrails are not broken and can be used effectively.

Take care of yourself.
- Do not run on uneven ground.
- Do not carry objects that obstruct your view of the ground.
- Exercise regularly to help retain your balance through your lifetime.
- Have your vision checked regularly.
- Be aware of any effects from medication that could reduce your alertness or balance.
- Make sure the treads on your work boots are not worn thin.

What are some tasks that have fall-related risks at this worksite?
(For example: icy walkways, uneven ground, poor lighting)

_____

_____

_____

_____

_____

_____

How can this worksite be changed to reduce fall-related risk?
(For example: deicing walkways, removing loose debris, designating a pathway, keeping equipment steps free of mud)

_____

_____

_____

_____

_____

_____

What can we do to reduce falls?
(For example: exercise, vision checks, maintaining three-point contact, adding lighting in low light areas)

_____

_____

Topic: Falls

| Topic: Falls | # Age Awareness Activity |

Areas of Risk at Your Site

Slips, trips, and falls can arise from a variety of sources that include environmental and personal factors. Flooring surface, slippery surfaces, boot design, and individual factors all contribute to the risk of falling. This exercise allows you to begin to identify these risks at your site so that you can work to correct them.

Equipment Needed

- Copies of the Mapping Risk Activity Form (see next page)
- Pencils or pens

Procedure

1. Hand out the Mapping Risk Activity Forms.
2. Ask trainees to draw a map of their work areas.
3. After trainees have completed their drawings, have them mark three places where there is risk for a slip, trip, or fall.
4. Lead a discussion where high-risk areas are shared.
5. Have trainees write down recommendations for lowering the risk in each of the places they marked on their maps.
6. Lead a discussion on the changes that could reduce risks. Be prepared to discuss how changes can become formal suggestions or be informally implemented.

Why Is This important?

Slips, trips, and falls are a leading cause of workplace injury. This activity encourages workers to think about the risks where they work. It may result in worksite changes that reduce risks, and it will definitely result in improved risk awareness.

For More Information

MSHA – Fall Hazards: http://www.msha.gov/smallmineoffice/toolbox/week9.pdf

Lehtola CJ, Becker WJ, Brown CM [2001]. Preventing injuries from slips, trips, and falls. Circular 869. Gainesville, FL: University of Florida, Agricultural and Biological Engineering Department, Institute of Food and Agricultural Sciences.

Mapping Risks

Draw a map of your work area in the box below. If you routinely work in more than one location, choose one for this activity.

Mark three (3) areas on your drawing where there is risk of a slip, trip, or fall.

List ways to reduce the risks you marked.

1.

2.

3.

Topic: Falls

| Topic: Falls | Age Awareness Resources |

Additional Resources

MSHA – New Technologies for Accident Prevention:
http://www.msha.gov/accident_prevention/newtechnologies/newtechnologies.asp

Lehtola CJ, Becker WJ, Brown CM [2001]. Preventing injuries from slips, trips, and falls. Circular 869. Gainesville, FL: University of Florida, Agricultural and Biological Engineering Department, Institute of Food and Agricultural Sciences.

OSHA – Safety and Health Topics, Fall Protection: http://www.osha.gov/SLTC/fallprotection/

Liberty Mutual – Slips, Trips, and Falls Publications:
http://www.libertymutual.com/omapps/ContentServer?cid=1029415782023&pagename=ResearchCenter%2FPage%2FStandardOrange&c=Page

MSHA – Fall Hazards: http://www.msha.gov/smallmineoffice/toolbox/week9.pdf

MSHA – Mounting and Dismounting Equipment:
http://www.msha.gov/stakeholderbp/BestPractices/MountDismount.pdf

Relevant Federal Regulations (intended as a guide, not a comprehensive list)
Title 30 CFR: Mine Safety and Health Administration (MSHA)

77.205 Travelways at surface installations: http://www.msha.gov/30cfr/77.205.htm

77.1710 Protective clothing; requirements: http://www.msha.gov/30cfr/77.1710.htm

56.11009 Walkways along conveyors: http://www.msha.gov/30cfr/56.11009.htm

56.11016 Snow and ice on walkways and travelways: http://www.msha.gov/30cfr/56.11016.htm

Title 29 CFR: Occupational Safety and Health Administration (OSHA)

1910.22 Walking-Working Surfaces:
http://www.osha.gov/pls/oshaweb/owadisp.show_document?p_table=STANDARDS&p_id=9714

1910.145(c)(2) Specifications for accident prevention signs and tags:
http://www.osha.gov/pls/oshaweb/owadisp.show_document?p_table=STANDARDS&p_id=9794

1910.23 Walking-Working Surfaces:
http://www.osha.gov/pls/oshaweb/owadisp.show_document?p_table=STANDARDS&p_id=9715

Topic:
Risky Tasks

Age Awareness Activity

Identifying High-Risk Tasks

This exercise provides the opportunity to evaluate the risk of an injury for a work task scenario. Several scenarios describe work tasks in various environments. Trainees will identify potential risk factors that may be affected by age-related changes and suggest changes that could reduce the risk of injury. You may want to tailor the scenarios or add an additional scenario using examples from your own worksite.

Equipment Needed

The scenarios provided can be read to the group, or copies can be given to individuals or small groups.

Procedure

1. Explain that this activity is not a test. It is a way to assess what everyone is currently thinking about age and work issues.
2. Have trainees read the scenarios (or read the scenarios to them).
3. Ask trainees (or small groups) to answer the questions that follow the scenarios.
4. Discuss the answers. See the discussion notes later in this section.
5. Ask trainees if similar situations exist at their workplace.

Why Is This Important?

Knowing about physical age-related changes and their relationships to the job is the beginning. This activity moves trainees to the next level by providing practice in assessing worksites and tasks where multiple potential hazards exist. This will help prepare workers to evaluate their own work areas.

Creating your own scenarios with local examples will make this exercise even more relevant to your trainees. You could also ask small groups of trainees to create scenarios for each other to review and assess.

NOTE: After this activity, it would be beneficial to have the trainees complete the Introduction Activity Form again. Look for differences from the first time the trainees completed the form. Any changes in the answers could be discussed with the trainees and management as a sign of success for the training. Include in this discussion the answers to the last question and how the suggestions have been or could be implemented.

Identifying High-Risk Tasks Exercise

Scenario 1: M. J.'s Ladder Carry

M. J. Smith is a mechanic who has worked in a maintenance shop at a nonmetal surface mine for 30 years. One busy day, M. J. was happy to be working on one of the benches located at the back of the shop because it was about 30 °F outside and had been snowing all day. During the last hour of the shift, a call came in to take a ladder assembly (approximately 30 pounds) to another building where the supervisor was working on a loader. The ladder was stored on the ground. To get to this location, M. J. had to walk out of the shop, across the road, through a small gravel parking lot, and into the building where the loader was parked.

1. What risks does M. J. face?
2. Are any of the risks increased for older workers? If yes, why?
3. How can the risks be reduced?

Scenario 2: Dozer Operation

T. R. Jones is a 52-year-old dozer operator who typically works a 12-hour rotating shift at a surface coal mine. During the latter part of the shift at about sunset, another dozer operator started working in the area. The other dozer moves closer and its backup alarm beeps. At the same time, T. R. hears on the radio that a contractor is entering the area in a small black pickup truck.

1. What risks does T. R. face?
2. Are any of the risks increased for older workers? If yes, why?
3. How can the risks be reduced?

Scenario 3: Supply Pickup

D. L. Bridges is a shuttle car operator at an underground coal mine (approximate seam height is 60 inches) who just turned 49 years old. At the beginning of a shift, D. L. is sent alone to a few crosscuts outby the dinner hole to get supplies (roof bolts, hydraulic oil, rock dust, etc.). The supplies are stored on the ground in a fairly wet place. There is no lighting in the area.

1. What risks does D. L. face?
2. Are any of the risks increased for older workers? If yes, why?
3. How can the risks be reduced?

Topic: Risky Tasks

INSTRUCTOR'S DISCUSSION NOTES

Scenario 1: M. J.'s Ladder Carry

M. J. Smith is a mechanic who has worked in a maintenance shop at a nonmetal surface mine for 30 years. One busy day, M. J. was happy to be working on one of the benches located at the back of the shop because it was about 30 °F outside and had been snowing all day. During the last hour of the shift, a call came in to take a ladder assembly (approximately 30 pounds) to another building where the supervisor was working on a loader. The ladder was stored on the ground. To get to this location, M. J. had to walk out of the shop, across the road, through a small gravel parking lot, and into the building where the loader was parked.

1. What risks does M. J. face?
    - With the transition from bright to dark, the worker may not be able to see any slip, trip, or fall hazards.
    - At the end of the shift, the worker may be tired and fatigued.
    - The weight of the equipment is heavy and may cause an overexertion injury.
    - Carrying such a large object may obstruct the worker's view or cause him/her to be distracted, which may affect the ability to pay attention to where he/she is walking.
    - If the ground had any uneven areas, the worker could trip or fall.
    - Given the temperature and weather conditions, there could be the potential for ice formation on the parking lot, which could cause the worker to slip.

2. Are any of the risks increased for older workers? If yes, why?
    - Vision changes when going from light to dark due to increased time to adapt to the dark
    - Lifting due to reduced strength, cumulative damage
    - Possibility of a slip or fall, because with age it may be harder to recover from a slip

3. How can the risks be reduced?
    - Provide a cart so that the employee does not have to carry the equipment.
    - Improve lighting in the building, or make sure the employee does not rush through areas that have extreme changes in lighting.
    - Place the equipment in a gator so that the employee can ride to the other part of the site.
    - Provide for adequate walkways that are free from debris, uneven areas, and ice.
    - Periodically check to see that treads on work boots are not worn out.

Topic: Risky Tasks

INSTRUCTOR'S DISCUSSION NOTES

Scenario 2: Dozer Operation

T. R. Jones is a 52-year-old dozer operator who typically works a 12-hour rotating shift at a surface coal mine. During the latter part of the shift at about sunset, another dozer operator started working in the area. The other dozer moves closer and its backup alarm beeps. At the same time, T. R. hears on the radio that a contractor is entering the area in a small black pickup truck.

1. What risks does T. R. face?
   - At the end of the shift, the worker may be tired and fatigued.
   - There may be glare due to the time of day, making it harder to see the dark truck.
   - The operator is distracted by both alarms or may not hear both alarms. Thus, the operator may not move his/her loader out of the way in time.
   - Being distracted, the worker might not sound his/her horn to alert the other loader.

2. Are any of the risks increased for older workers? If yes, why?
   - Fatigue from end of shift (older workers often have trouble adjusting to rotating shift work)
   - Possible problems with glare due to increased glare sensitivity with age
   - Divided attention task because attending to more than one task is more difficult for older workers
   - Hearing—if both alarms are similar and it is difficult to distinguish between them

3. How can the risks be reduced?
   - Arrange the tasks so that the dozers don't have to be so close to one another.
   - Put into place safety procedures limiting long radio conversations while driving.
   - Require vehicles to be well-lit when driving on-site.
   - Provide redundant signals.

Topic: Risky Tasks

INSTRUCTOR'S DISCUSSION NOTES

Scenario 3: Supply Pickup

D. L. Bridges is a shuttle car operator at an underground coal mine (approximate seam height is 60 inches) who just turned 49 years old. At the beginning of a shift, D. L. is sent alone to a few crosscuts outby the dinner hole to get supplies (roof bolts, hydraulic oil, rock dust, etc.). The supplies are stored on the ground in a fairly wet place. There is no lighting in the area.

1. What risks does D. L. face?
   - Muscle strain due to loading items alone
   - Muscle strain due to lifting and bending
   - Possible slip due to the wet environment and poor lighting

2. Are any of the risks increased for older workers? If yes, why?
   - Since the worker had just started the shift, his/her muscles were not warmed up. This might add to the risk of muscle strain.
   - Muscle strain due to leaning over to get supplies off the ground may occur since older workers may have a reduction in flexibility.
   - Unevenness in ground or mud might cause more of a fall hazard for older workers.
   - Poor lighting might also lead to a slip or fall.

3. How can the risks be reduced?
   - Send two workers to lift items, not just one.
   - If possible, try to place items so that lifting occurs at waist height.
   - Warm up by lifting slowly at first and then ramping up to normal lifting frequency.
   - Keep areas by supplies free from debris that could cause a tripping hazard.

Topic: Risky Tasks

| Topic: Records | Age Awareness Training |

GENERAL LEVEL TRAINING EVALUATION

Leader:

Date/shift of this training session: Number of trainees:

Module(s) included:

- ☐ Introduction
- ☐ Vision
- ☐ Hearing
- ☐ Attention and Memory
- ☐ Musculoskeletal System
- ☐ Lower Back
- ☐ Work Capacity
- ☐ Slips and Falls
- ☐ Risky Tasks

Material that was presented or used:

- ☐ Newsletter
    How was it distributed? _____
- ☐ Safety Talk Guide
- ☐ Activity 1 _____
- ☐ Activity 2 _____
- ☐ Additional Resources

Was this topic relevant to your work group?   ☐ YES   ☐ NO
If NO, please describe why not.

Should additional training be done on this topic in the future?   ☐ YES   ☐ NO
If YES, please explain specific areas that would be useful.

Were there any specific recommendations for workplace changes from this topic?
                                                  ☐ YES   ☐ NO
If YES, are explanations for these changes attached?   ☐ YES   ☐ NO

| Topic: Records | Age Awareness Training |

LEADER LEVEL TRAINING EVALUATION

Date: Instructor(s):

Number of leaders:
Description of leader group (e.g., all night-shift supervisors):

PowerPoint slides
- ☐ Used the slides as given
- ☐ Modified the slides or deleted some
- ☐ Didn't use the slides

Presentation notes
- ☐ Used the notes during the presentation
- ☐ Read the notes while preparing my presentation
- ☐ Didn't use the notes

What were the general reactions of the leaders to this session?

What should be changed for future sessions?

What other employees should receive this training?

References — Age Awareness Training

References for Introduction

MSHA [2005]. Accident, illness and injury and employment self-extracting files (part 50 data), 1992–2002. Denver, CO: U.S. Department of Labor, Mine Safety and Health Administration, Office of Injury and Employment Information. [http://www.msha.gov/STATS/PART50/p50y2k/p50y2k.HTM]. Date accessed: July 2005.

References for Leader Level Training

Barth MC [1997]. Older workers: perception and reality. Presentation to the U.S. Special Committee on Aging Forum on July 25, 1997.

Hedden T, Gabrieli JDE [2004]. Insights into the aging mind: a view from cognitive neuroscience. Nat Rev Neurosci *5*(2):87–96.

Krause N, Ragland DR, Risher JM, Syme SL [1998]. Psychosocial job factors, physical workload, and incidence of work-related spinal injury: a 5-year prospective study of urban transit operators. Spine *23*(23):2507–2516.

McNaught W, Barth M [1992]. Are older workers "good buys": a case study of Days Inns of America. Sloan Manag Rev *Spring*:53–63.

MSHA [2005]. Accident, illness and injury and employment self-extracting files (part 50 data), 1992–2002. Denver, CO: U.S. Department of Labor, Mine Safety and Health Administration, Office of Injury and Employment Information. [http://www.msha.gov/STATS/PART50/p50y2k/p50y2k.HTM]. Date accessed: July 2005.

Salminen S [2004]. Have young workers more injuries than older ones? An international literature review. J Safety Res *35*(5):513–521.

Salthouse TA [1984]. Effects of age and skill in typing. J Exp Psychol Gen *113*(3):345–371.

Sorenson G, Stoddard A, Ockene JK, Hunt MK, Youngstrom R [1996]. Worker participation in an integrated health promotion/health protection program: results from the WellWorks project. Health Educ Q *23*(2):191–203.

Toossi M [2002]. A century of change: the U.S. labor force, 1950–2050. Mon Labor Rev *125*(5):15–28.

Walsh DW, Jennings SE, Mangione T, Merrigan DM [1991]. Health promotion versus health protection? Employees' perceptions and concerns. J Public Health Policy *12*:148–164.

Warr R [1994]. Age and job performance. In: Snel J, Cremer R, eds. Work and aging: a European prospective. Taylor & Francis, pp. 309–322.

References for Vision Content

Buch ER, Young S, Contreras-Vidal JL [2003]. Visuomotor adaptation in normal aging. Learn Mem *10*(1):55–63.

Charness N, Schaie KW, eds. [2003]. Impact of technology on successful aging. New York: Springer Publishing Co., Inc.

Fisk AD, Rogers WA, Charness N, Czaja SJ, Sharit J [2004]. Designing for older adults: principles and creative human factors approaches. Boca Raton, FL: CRC Press.

La Haye P Jr., Sustello R [2001]. Safety eyewear and the older worker. Occup Health Saf *70*(10):38–51.

Takeichi K, Sagawa K, Kuchinomachi Y, Kanaya S [1997]. Color temperature of light source and discomfort glare of elderly people. In: Aging and Ergonomics, 13th Triennial Congress of the International Ergonomics Association (Tampere, Finland).

References for Hearing Content

Charness N, Schaie KW, eds. [2003]. Impact of technology on successful aging. New York: Springer Publishing Co., Inc.

Fisk AD, Rogers WA, Charness N, Czaja SJ, Sharit J [2004]. Designing for older adults: principles and creative human factors approaches. Boca Raton, FL: CRC Press.

Hearing Loss Association of America [2008]. Self-assessment hearing test. [http://www.hearingloss.org/learn/self-assessment.asp]. Date accessed: January 2008.

National Research Council [2004]. Health and safety needs of older workers. Washington, DC: National Academies Press.

Washington University Hospitals [2008]. Department of otolaryngology: how does the ear work? [http://wuphysicians.wustl.edu/dept.asp?pageID=14&ID=8]. Date accessed: January 2008.

References for Attention and Memory Content

Barber CE [2001]. Age-related changes in memory. [http://www.ext.colostate.edu/pubs/consumer/10243.html]. Date accessed: January 2008.

Charness N, Schaie KW, eds. [2003]. Impact of technology on successful aging. New York: Springer Publishing Co., Inc.

Fisk AD, Rogers WA, Charness N, Czaja SJ, Sharit J [2004]. Designing for older adults: principles and creative human factors approaches. Boca Raton, FL: CRC Press.

National Research Council [2004]. Health and safety needs of older workers. Washington, DC: National Academies Press.

References for Musculoskeletal Content

Brooks GA, Fahey TD [1985]. Exercise physiology: human bioenergetics and its applications. Chapter 31: Aging. New York: Macmillan Publishing Co., pp. 683–699.

Ergoweb [2008]. OSHA general duty clause. [http://www.ergoweb.com/resources/reference/guidelines/oshageneraldutyclause.cfm]. Date accessed: January 2008.

National Research Council and Institute of Medicine [2001]. Musculoskeletal disorders and the workplace: low back and upper extremities. Washington, DC: National Academies Press.

NIH consensus development panel on optimal calcium intake [1994]. JAMA *272*(24):1942–1948.

Putz-Anderson V, ed. [1988]. Cumulative trauma disorders: a manual for musculoskeletal disease of the upper limbs. London: Taylor and Francis.

References for Lower Back Content

Adams MA [1995]. Spine update – Mechanical testing of the spine: an appraisal of methodology, results, and conclusions. Spine *20*(19):2151–2156.

Adams MA, Freeman BJC, Morrison HP, Nelson IW, Dolan P [2000]. Mechanical initiation of intervertebral disc degeneration. Spine *25*(13):1625–1636.

Bogduk N [1997]. Clinical anatomy of the lumbar spine and sacrum. 3rd ed. New York: Churchill Livingstone.

Brinckmann P, Biggemann M, Hilweg D [1988]. Fatigue fracture of human lumbar vertebrae. Clin Biomech *3*(Suppl 1):1–23.

Ergoweb [2008]. OSHA general duty clause. [http://www.ergoweb.com/resources/reference/guidelines/oshageneraldutyclause.cfm]. Date accessed: January 2008.

Gallagher S, Marras WS, Litsky AS, Burr D [2005]. Torso flexion loads and the fatigue failure of human lumbosacral motion segments. Spine *30*(20):2265–2273.

Solomonow M [2005]. Ligaments: a source of work-related musculoskeletal disorders. J Electromyogr Kinesiol *14*(1):49–60.

References for Work Capacity Content

Astrand PO, Rodahl K [1977]. Textbook of work physiology: physiological bases of exercise. New York: McGraw-Hill.

Ilmarinen JE [2001]. Aging workers. Occup Environ Med *58*(8):546–552.

National Research Council [2004]. Health and safety needs of older workers. Washington, DC: National Academies Press.

Sharkey BJ [1979]. Physiology of fitness. Champaign, IL: Human Kinetics Publishers.

Sharkey BJ [1997]. Fitness and work capacity. 2nd ed. Missoula, MT: U.S. Department of Agriculture, Forest Service, Missoula Technology and Development Center, NFES 1596.

References for Slips, Trips, and Falls Content

Bell JL, Gardner LI, Landsittel DP [2000]. Slip and fall-related injuries in relation to environmental cold and work location in above-ground coal mining operations. Am J Ind Med *38*(1):40–48.

CDC, National Center for Injury Prevention and Control [2008]. Preventing falls among older adults. [http://www.cdc.gov/ncipc/pub-res/toolkit/toolkit.htm]. Date accessed: January 2008.

Chang W-R [2004]. Assessing floor slipperiness in fast-food restaurants in Taiwan using objective and subjective measures. Appl Ergon *35*(4):401–408.

Courtney TK, Sorock GS, Manning DP, Collins JW, Holbein-Jenny MA [2001]. Occupational slip, trip, and fall-related injuries: can the contribution of slipperiness be isolated? Ergonomics *44*(13):1118–1137.

Fisk AD, Rogers WA, Charness N, Czaja SJ, Sharit J [2004]. Designing for older adults: principles and creative human factors approaches. Boca Raton, FL: CRC Press.

Haslam R, Stubbs D, eds. [2006]. Understanding and preventing falls. Boca Raton, FL: CRC Press LLC.

U.S. Bureau of Labor Statistics [2003]. Fatal occupational injuries by event or exposure and age: all United States, 2003. [http://www.bls.gov/iif/oshwc/cfoi/cftb0194.pdf]. Date accessed: January 2008.

U.S. Bureau of Labor Statistics [2003]. Nonfatal cases involving days away from work: selected characteristics. [http://data.bls.gov/cgi-bin/surveymost?ch]. Date accessed: January 2008.

Walton M [1999]. Graying, not falling. Occup Health Saf *68*(4):85–87, 92.

| Glossary of Terms | # Age Awareness Training |

Acute injury
    Acute injuries happen instantly, such as fractures, cuts, and bruises.

Awkward posture
    A body position where one or more joints of the body are placed at an uncomfortable, non-neutral angle, often at or near the limit of its range of motion. Excessive torso bending, reaching away from the body, bending the neck, or reaching above shoulder height are examples of awkward postures. Whenever possible, arrange the work station or work processes to allow employees to work from a comfortable, neutral posture.

Carpal tunnel syndrome
    A disorder causing pain, numbness, and tingling of the hand caused by compression of the median nerve as it passes through the carpal tunnel in the wrist.

Cumulative injury
    Cumulative injuries develop as a result of repeated loading of body tissues over time. Such injuries may result in sprains/strains, herniated discs, tendonitis, and carpal tunnel syndrome.

Cumulative trauma disorders (CTDs)
    Injuries and illnesses that result from exposure to repeated stresses over a period of time. They affect soft tissues and bones of the musculoskeletal system and/or nerves and blood vessels servicing the musculoskeletal system.

Discomfort
    Discomfort can involve mental or physical distress. Examples of physical distress include aches and pains your body is experiencing. Examples of mental distress include loss of a loved one, pressure to perform at work, or lack of sleep.

Ergonomics
    Ergonomics is the scientific study of people at work. Ergonomics seeks to match the physical and cognitive requirements of the job to the abilities of the worker. This is achieved by designing workplaces, environments, job tasks, equipment, and processes to suit the worker's abilities.

Fatigue
    Fatigue is a complex problem that has both physical and mental effects. Local fatigue occurs when muscles are not able to recover and is a function of how hard and how long a person works. Local fatigue results in an inability to continue to perform work at the same rate. General fatigue has mental effects such as diminished alertness and reduced perceptual capacity, which can lead to faulty judgment, increased error rates, and mishaps.

Force
    The amount of physical effort a person must exert to perform a task.

Hand-arm vibration
    Vibration (generally from a hand tool) that is transmitted through the hand, often leading to the development of upper-extremity CTDs.

Long-term memory
    The part of memory that stores items that are events in time, knowledge about words or concepts, or ways to do certain tasks, such as standard operating procedures.

Mean
    The mean of a set of numbers is found by dividing the sum of the data by the number of entries. It is also called the average.

Median
    The median of a set of numbers is the midpoint, or the value at which half of the numbers are smaller and half of the numbers are larger.

Musculoskeletal disorders (MSDs)
    Illnesses and injuries that affect the muscles, ligaments, tendons, cartilage, and bones in the body and/or their associated nerves and blood vessels.

Musculoskeletal system
    The bones, joints, and surrounding soft tissue such as muscles, ligaments, and tendons.

Neutral posture
    A comfortable working posture that reduces the risk of musculoskeletal disorders. An ideal posture for the body would be standing with arms at your sides, elbows bent, wrists straight, and eyes looking straight ahead.

Personal protective equipment (PPE)
    Gloves, kneepads, and other equipment that may help reduce hazards until other controls can be implemented or that supplement existing controls.

Presbycusis
    Age-related hearing loss.

Presbyopia
    Age-related changes to the eyes, specifically, the loss of the ability to focus the eyes on near objects.

Repetitiveness
    Occurs when performing the same motions repeatedly. The severity of risk depends on the frequency of repetition, speed of the movement or action, the number of muscle groups involved, and the required force.

Risk factors
    An aspect of the job that increases the chance of getting a work-related MSD.

Short-term (working) memory
    The part of memory that stores a limited amount of information for a short period of time, such as what you did earlier in the day.

Sprain
    Overstretching or overexertion of a ligament that results in a tear or rupture of the ligament.

Strain
    Overstretching or overexertion of a muscle or tendon.

Tendonitis
    Inflammation of the tendon.

Tenosynovitis
    Inflammation of the sheath surrounding the tendon.

Whole-body vibration
    Exposure of the whole body to vibration (generally transmitted through the feet/buttocks when riding in a vehicle).

Work capacity
    The maximum amount of effort a person can put forth.